The Yellow-Legged Asian Hornet

A Handbook

First published 2019 as The Asian Hornet Handbook
Second edition with major changes 2024
2.3 (updated February 2025)
Psocid Press

ISBN-13
978-1-9160871-1-8

Illustrations and layout by the author. If you have any insect book projects, contact me!

Psocid (pronounced 'so-sid') Press runs on tea and jaffa cakes.

For the insects

Contents

Control

A note about the common name

In China, *Vespa velutina* is known as the yellow-legged hornet (sometimes the black hornet) and, although there are other hornets with yellow legs, the bright yellow legs in contrast with the largely black body are a very distinctive characteristic. Having a common name that reflects the actual appearance of the hornet is an important advantage when trying to teach people to recognise it, and therefore be able to report it.

There are around 22 species of hornet in the world, all originating in Asia. Our native hornet, the European hornet *Vespa crabro,* is actually found all the way across the great land mass from Europe to Japan (it was also introduced into North America, but that's another story).

So, is it helpful to talk about 'the Asian hornet' when all hornets evolved in Asia and there is ample room for confusion with the 'Asian **giant** hornet' (preferred common name 'northern giant hornet'), *Vespa mandarinia*? What happens when another hornet arrives in Europe? Already, along with *Vespa crabro*, two other hornets, *Vespa bicolor* and *Vespa orientalis* are established in Europe, the latter heading north and now found in northern Italy, Romania and France (Zucca et al. 2023). *Vespa soror* ('southern giant hornet') appeared in northern Spain in 2022 and 2023, leading to a major monitoring campaign. Climate change and global trade will see hornets expanding their territory worldwide.

The scientific literature, although mainly using its scientific name, has shifted over the last few years from Asian hornet to yellow-legged hornet or YLH (sometimes Y-L hornet) as the common name used in articles.

Throughout this book I refer to this hornet by its scientific name, *Vespa velutina* (*V. velutina* for short), or the yellow-legged hornet (YLH).

Introduction

The yellow-legged hornet (*Vespa velutina*) arrived in France from China around 2004 and has since established or been recorded in 16 other European countries. In 2023, there was a large influx into southern England. This hornet is a voracious hunter of insects, with a particular taste for honey bees, and attacks humans if its nest is approached or disturbed.

In France and Spain, it is becoming more common to find densities of nests reaching 15 or more per square kilometre in urban or semi-urban environments. To stop yellow-legged hornets from establishing in the UK, the first action must be to raise awareness, so that people can recognise them and are motivated to report them. Please get involved with this task.

I want this book to be a primer for those who perhaps don't know anything about hornets but who are inventive. We desperately need traps that are more selective, keeping by-catch of native insects to an absolute minimum, and we need cheap, fast, efficient and ecologically friendly ways of finding and destroying nests. Everything I could find out about these hornets has gone into this book. Once you understand them, perhaps you can see a way they can be controlled.

There has been much research since I published the first edition in 2019. This updated and expanded version is divided into three parts: it starts with **Biology**, looking in detail at what we know so far of their life history and ecology. I give no apology for the depth of this section; only by understanding their natural history can we hope to control them without upsetting the ecological balance. The **Context** gives information on the story as it has unfolded in Europe (including the UK); and in **Control**, national strategies, tracking, destruction of nests, trapping and other methods are discussed, as well as how to deal with them as a beekeeper.

I have tried to keep the language easy to follow, but there are quite a few specific terms that are important, so I have provided a Glossary (with an illustration of hornet body parts). For those who want to go deeper, there is a full reference list (most articles can be found online). This version has a few updates, and an expanded index.

Sarah Bunker, February 2025

Part I
Biology

Identification

Yellow-legged hornets are straightforward to identify, and if you are familiar with our native European hornet (*Vespa crabro*) you will quickly see how different they are. On the following pages we look at their key identification features and insects that are sometimes confused with them. For meanings of technical terms and an illustration of parts of the hornet body, see the Glossary at the end of the book.

The scientific name for the yellow-legged or Asian hornet is *Vespa velutina*. In fact, the full name for the colour form that has arrived in Europe is *Vespa velutina nigrithorax*. *Vespa* is the scientific name for hornets generally (the genus), and *velutina* (the species name) means 'velvety'. The last name (*nigrithorax*) refers to the black thorax. In Asia, the original home of this hornet, there are 12 colour forms, with largely distinct territories. *Vespa velutina nigrithorax* is found in central and eastern China as well as in the Himalayan mountains.

The yellow-legged hornet ranges from around 25 mm to 30 mm in length (head to tail), which is **smaller** than our native European hornet. At the beginning of the season, the workers produced are smaller, and those produced later in the year are *usually* larger, due to being better fed as larvae, and having larger cells in which to grow. The YLH is mainly a velvety black or very dark brown, with a thin yellow band on the abdomen close to the thorax and a wider orange-yellow band closer to its tail. It also has an orange face (the head is black from above) and dark smoky-brown wings. Like other hornets and wasps, the wings are folded longitudinally when not in use, and form a distinctive 'V' shape when they are resting or walking. The top halves of its legs are a dark brown-black and the bottom halves are yellow. It has long brown/black antennae. From underneath (if you see one in a transparent container), the body has a bit more yellow and is not as distinctive (see illustration). They fly during the day and into dusk and dawn, but do not fly when it is fully dark (unlike the European hornet) except when disturbed at night, for example if a nest is removed after dark.

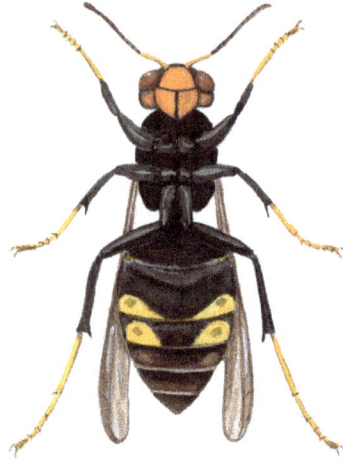

This page: the yellow-legged hornet, Vespa velutina nigrithorax.

Above: the underside (ventral side). Note the orange face.

Left: the wings are folded longitudinally and held over the back like this, or in more of an open 'V' when walking.

Dried specimen. With its wings stretched out you can see that there are two pairs. The sting is also visible. The colours of the dried specimen are duller than in a live insect. Photo by Didier Descouens (CC BY-SA 4.0).

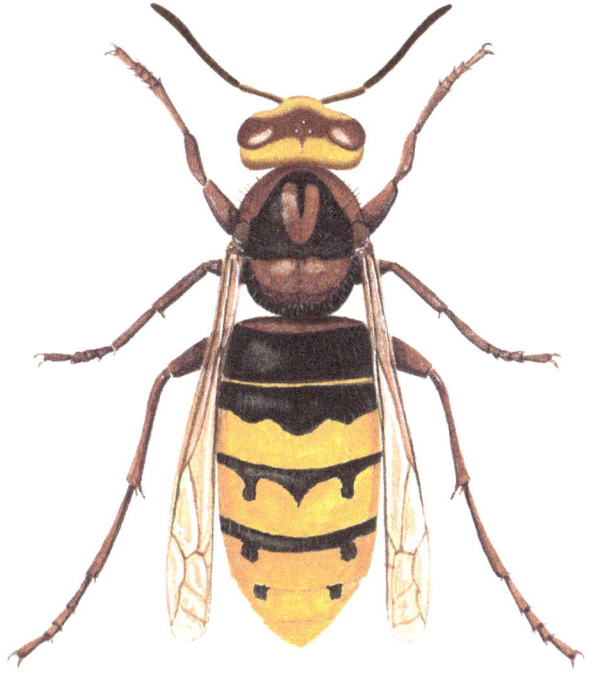

This page: our native European hornet, Vespa crabro gribodoi. *The one on the **left** with long sweeping antennae is a male. The markings on the abdomen are variable.*

What can yellow-legged hornets be confused with?

European hornet (*Vespa crabro*)

Our native European hornet, *Vespa crabro*, is **larger** than *Vespa velutina* (it is around 30–35 mm long) and looks like a very large wasp. It is the only other hornet seen in northern Europe. *Vespa crabro* is bulkier than the yellow-legged hornet, with a large and sometimes drooping abdomen. The markings on its abdomen are a deep yellow, and chestnut brown to black; the pattern is very distinctive: the darker marks form shapes that look just like the lugs on jigsaw pieces. It also has chestnut (red-brown) legs and a black and brown thorax. It has a yellow face and brown to black antennae. The European hornet is often active at night, sometimes flying into rooms through open windows on warm evenings, attracted to the light.

Although our native hornet is big, it is no more likely to sting people than an ordinary wasp, but with its longer sting and larger size, it can deliver more venom and a painful sting.

Vespa crabro builds quite large paper nests, but they are usually under cover in a tree-hole, wall or loft space; occasionally underground. The colony (apart from the new generation of queens that hibernate after mating) dies off in the autumn, but the nest site (not the nest itself) may be used again the following year. European hornet nests have a large entrance at the bottom, whereas only embryo and primary nests of yellow-legged hornets have a (small) entrance at the bottom. Colonies typically have fewer individuals than yellow-legged hornets. *Vespa crabro* nests may contain 200-400 (rarely 1,000) workers at their peak, while *Vespa velutina* nests may have an average of 400-500, but larger nests contain 1,000-2,000 (possibly more) workers at peak numbers in the autumn.

Horntail or Giant wood wasp (*Urocerus gigas*)

This spectacular insect is a type of sawfly. The females have what look like terrifyingly long stings, but these are in fact harmless ovipositors that they use to deposit their eggs in tree wood. They are quite a bit larger than hornets — the females get up to 45 mm long. They have a distinctive yellow and black banded abdomen, yellow legs and long yellow antennae. They have yellow/golden patches behind their black eyes. They are found in woods and sometimes emerge from construction timber on building sites and in timber yards.

Urocerus gigas, *the horntail, laying eggs. Photo by Holger Gröschl (CC BY-SA 2.0).*

Hornet mimic hover-fly (*Volucella zonaria*)

Volucella zonaria, hornet mimic hover-fly. Photo by Alvesgaspar (CC BY-SA 3.0).

As the name suggests, this is a type of fly that has evolved to look dangerous like a hornet, but this is a ruse — it has no sting and is completely harmless. They are generally a bit smaller (20 mm or less) than a yellow-legged hornet. The banding on the abdomen has much more yellow than the yellow-legged hornet and its legs are dark. When it is walking, it puts its wings over its back, but they aren't folded up like a wasp or hornet. The most distinctive difference between *Volucella zonaria* and the yellow-legged hornet is the antennae: the hover-fly has very small whisker-like antennae (barely visible in the photo), whereas hornets have large distinctive antennae.

Wasps

Technically, hornets are wasps: to entomologists, they are part of a bigger 'wasp' family. What we generally refer to as wasps in the British Isles are the smaller black-and-yellow insects that can be annoying in pub gardens and orchards. They are significantly smaller than the yellow-legged hornet (they are between 10 and 20 mm), and shinier. There are several species in this country; some build nests in protected, dark places, such as in the ground, in compost heaps and

Vespula vulgaris (male). Photo by Magne Flåten (CC BY-SA 3.0).

in building cavities, while others build exposed nests in bushes and small trees. The paper nest is roughly spherical and has a small entrance at the bottom. One thing that all these wasps have in common is some form of yellow marking on the thorax, whether spots or 'shoulder' stripes: the yellow-legged hornet has a completely black thorax.

Dolichovespula media. *Photo by PJT56 / Wikimedia Commons / CC BY-SA 4.0.*

Dolichovespula saxonica. *Photo by Sandy Rae (CC BY-SA 3.0).*

A note about the Asian *giant* hornet

The Asian (or Japanese) **giant** hornet (*Vespa mandarinia*), known now as the **northern giant hornet**, has not yet been found in Europe, but that doesn't stop newspapers and online articles using pictures of it when they should be using pictures of *Vespa velutina*. This is the world's largest hornet (around 45 mm long, and chunky) and its size and completely yellow head make it more dramatic than *Vespa velutina*. It was recently found, and eradicated from the north west of the USA.

Vespula rufa. *Photo by Richard Bartz, Munich aka Makro Freak (CC BY-SA 2.5).*

Vespa mandarinia japonica, *the Asian **giant** hornet (female). Photo by Yasunori Koide (CC BY-SA 4.0).*

Identifying different castes and sexes

Like other social hymenopterans, such as honey bees and wasps, hornets have three 'castes': queens and workers (both female), and males. The queen is usually the sole reproductive insect, while the female workers forage for food and nest material, feed larvae, build the nest, keep the temperature at the right level and defend the nest. As the year progresses, the queen keeps laying and the workers feed the larvae and expand the nest. At the peak of colony life, in the autumn, the sexuals are produced. They mate (ideally with sexuals from other colonies to avoid inbreeding) and the mated queens go on to found colonies of their own the following spring. For those who remove nests it is important to be able to identify the breeding queen (sometimes called the 'mother queen'), the workers, the males and the 'gynes' (virgin queens). Recognising these castes allows you to assess whether or not the nest has released sexuals.

Females (workers and queens) have pointed 'tails' (the tips of their abdomens), and stings. They have 10 segments to their antennae. Males have much blunter 'tails', even flat at the very tip, and on the underside, close to the tip of the tail are two small yellow spots (see photo). Males don't have stings. They have 11 segments to their antennae (the segments are difficult to count without a lot of magnification), and their antennae are usually more curved.

You are most likely to come across workers (perhaps from late May onwards in the UK: timings from Jersey), because they are busy with colony tasks, including foraging. They are present in the colony all through the season, once the queen has raised her first batch.

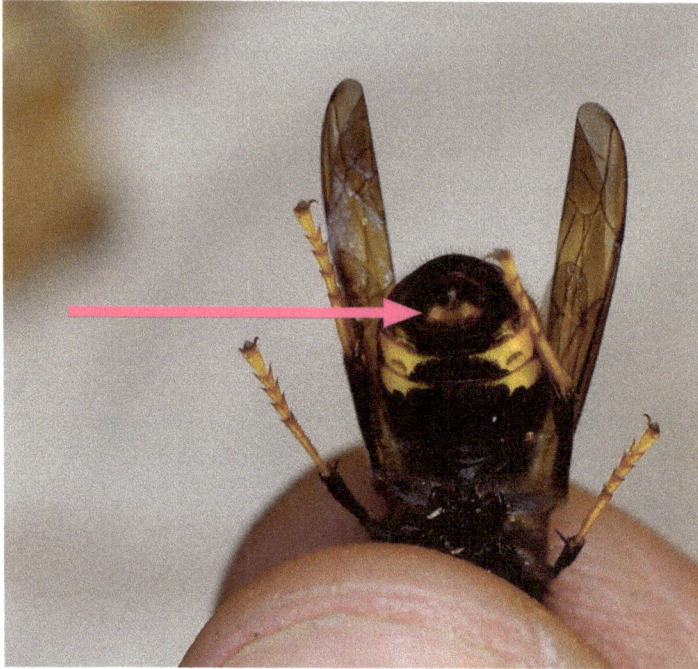

*Male yellow-legged hornet showing two yellow spots (**arrow**) which distinguish it from a female. Photo by John de Carteret.*

If you see a yellow-legged hornet between the middle of February and late May (timings from Jersey — it may be later if they become established in the UK), it is most likely to be a queen, because only the queen is expected to hibernate over winter in our climate (the workers and males die off), and once she builds an embryo nest it still takes another 50 days for the first workers to appear.

The first batch of workers are usually significantly smaller than later workers because they are not fed as well. So, early in the season, the colony queen can be picked out by her larger size. By the autumn, when workers are generally much larger, queens and workers are not so easy to distinguish. However, once the queen has spent a long time confined to the nest, she becomes shiny from constant attention by workers, who rub off her body hair (Martin 2017), making her easier to identify. In addition to size, the queen can be distinguished by having large fat bodies in her abdomen, as well as eggs.

Probably the trickiest distinction to make is between workers and the new virgin queens (**gynes**) produced in the autumn. They are both female and of similar sizes, but gynes are slightly larger and heavier. Because the gynes are not yet mated, they won't have eggs, and if they have not fed for long, they won't have much in the way of fat bodies. It's vital to be able to make this distinction because the appearance of gynes in the nest signals the imminent departure of sexuals. Once the sexuals have departed, you have lost the opportunity to destroy the offspring of that nest.

By sampling 2,744 females, Rome et al. (2015) found that the weights fell into two classes: gynes and workers, with those weighing less than 525 mg being workers and those above 665 mg being gynes (wet weight). Weights between these two were uncertain. Scales that can measure to the milligram are now available very cheaply.

Dead queen from a removed nest dissected to show eggs in abdomen. Photo by Ana Diéguez-Antón.

Kim and Choi (2021) used average head width to distinguish between gynes and workers: gyne head widths were 5.98 ± 0.03 mm; workers were 4.88 ± 0.11 mm. But it's important to note that they only used five individuals of each caste to get these data.

Probably the most accessible technique for distinguishing gynes from workers is to measure the mesoscutum width (MW: the middle part of the thorax, see illustration). Pérez-de-Heredia et al. (2017) found that, in a late season nest, 249 females fell into two classes of MW: workers with a mesoscutum width under 4.5 mm

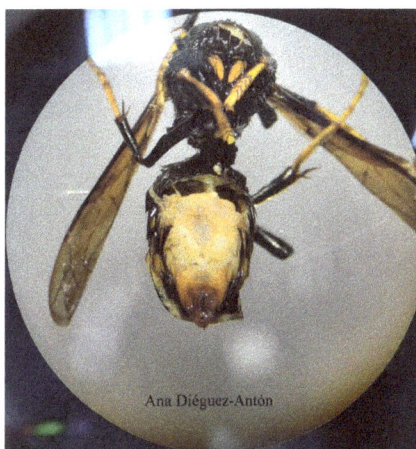

Dead queen from a removed nest dissected to show fat bodies in abdomen. Photo by Ana Diéguez-Antón.

and gynes with a mesoscutum width of 4.5 mm or above. Once again, there is some uncertainty (one gyne had an MW of only 4.48 mm), but it is a useful rule of thumb. MW can be measured using a reticle in the eyepiece of a microscope, or by using modified digital calipers (see photos).

Interestingly, it is possible to largely distinguish pupae by caste. The silk caps on the cells where the hornets are developing from larvae into adults are of different heights. Diéguez-Antón et al. (2022a) found that the tallest caps belonged to future queens (gynes), the next tallest belonged to workers, and the shortest belonged to males.

Pupal case sizes differing between future queens (yellow dots), *workers* (red dots), *and males* (blue dots). *Photograph from Dieguez-Anton et al. 2022a, with permission (CC BY 4.0).*

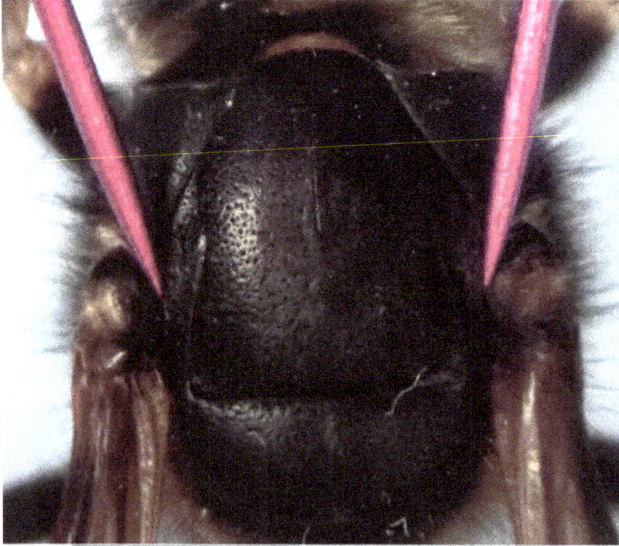

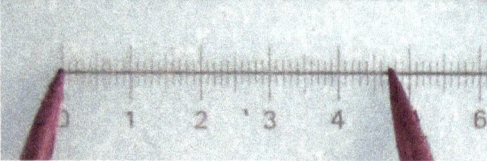

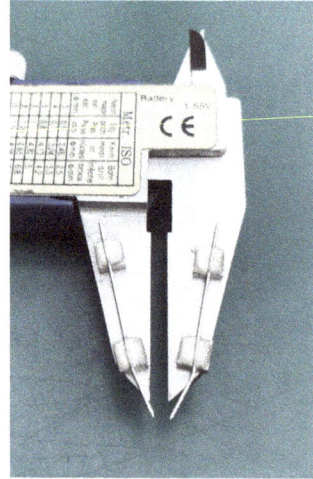

Left: *the mesoscutum width, MW (between needle points). Photo by Chris Isaacs.*
Above: *Chris Isaacs' digital calipers modified with bent pins to make measuring the MW easier.*

Life Cycle

Emergence from hibernation

The cycle starts when mated queens, also known as 'foundresses', come out of hibernation in the spring. They will have been hibernating in woodpiles, burrows, crevices, tree bark, soil, in decaying wood or under stones (Monceau et al. 2014a, Marris et al. 2011) and may hibernate singly in small cavities that they excavate for themselves (called a hibernaculum: Van Itterbeeck et al. 2021), or in clusters of two, three or four (Chauzat & Martin 2009, Poidatz 2017), in a characteristic pose with the wings tucked under the abdomen (Martin 2017). Hibernation sites tend to be cool, moist and dim, and mostly north-facing (Van Itterbeeck et al. 2021), and the cavities they occupy are often excavated in pale, light 'punky' wood that fungus has been through.

Two queens hibernating together in a hibernaculum. Photo by Dominique Soete.

In Jersey, although a few queens were found in February in 2022, general emergence seems to occur in late March, when the temperature has reached 12 ˚C. Certainly, none were trapped in France below 10 ˚C (Monceau et al. 2012).

13

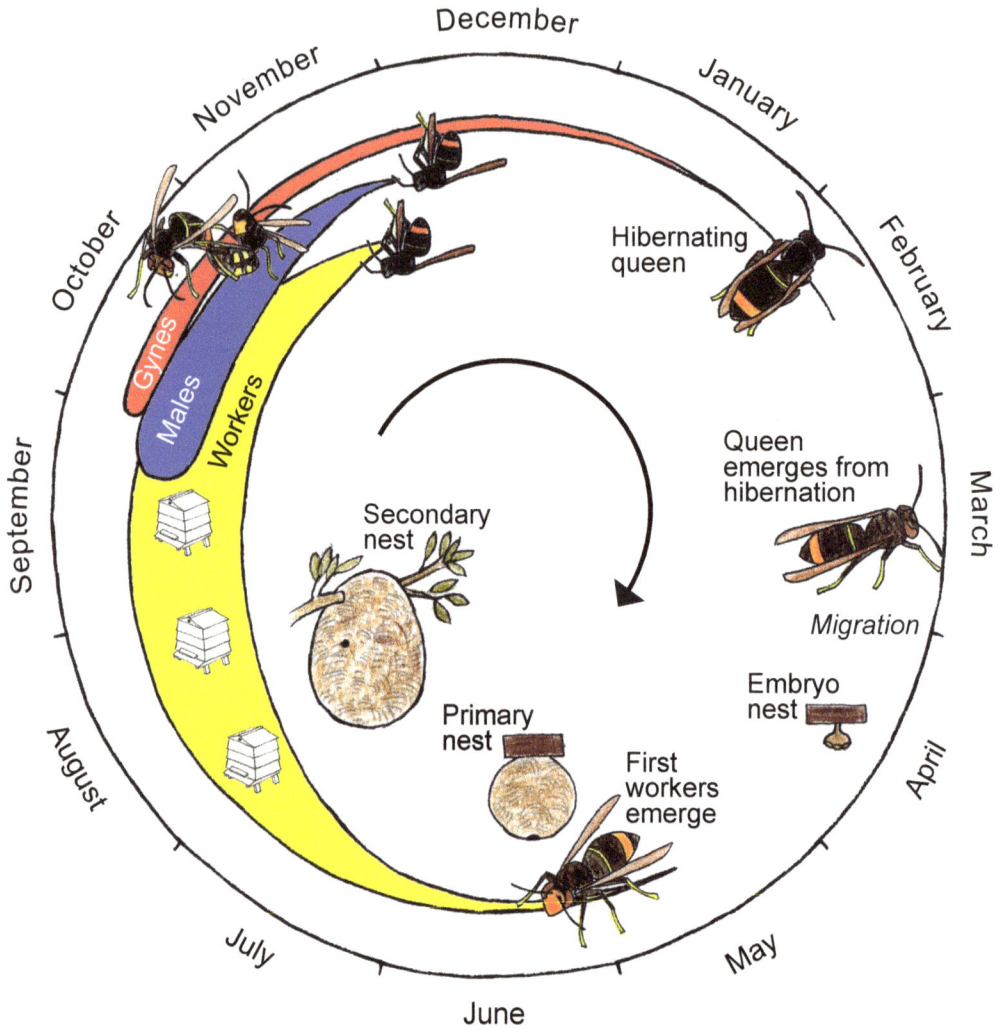

Life cycle of Vespa velutina. *The **yellow area** indicates the increase and decrease in numbers of workers. Towards the end of the season, the sexual stages are produced; first the males (**blue**) and then the females (**red**). The female sexual stages are called gynes and once fertilised will become next year's queens. The timings here are based on Jersey data.*

Once emerged, foundress queens rebuild their strength by feeding on nectar or tree sap. The most important food source for large-sized *Vespa* species in spring is the sap secreted from the trunks of oak in addition to those of beech, maple and willow (Matsuura & Yamane 1990). In Jersey and France, yellow-legged hornet queens have also been noticed feeding on spring-flowering camellias.

14

Migration

In other species of *Vespa*, post-hibernation migration of queens as small groups has been witnessed as swarms, or assumed from mark and recapture experiments when many marked queens have disappeared from an area (Matsuura & Yamane 1990). In *Vespa velutina*, it is assumed that some post-hibernation migration occurs, (see p 82), although some of the spread may be assisted by hitch-hiking on human transport. Lioy et al. (2019), looking at data for Liguria, northern Italy, found that 95% of nests were built within 1.4–6.2 km of previous nests, but the remaining 5% were widespread single nests. Thus, the majority of queens are only travelling a few kilometres from their hibernation site, while a few may travel 10 or 20 km. Rome et al. (2015) found, in preliminary flight-mill experiments, that YLH queens were able to fly more than 40 km per day, although flight mills are a long way from natural flight, so this may not occur in the real world (see 'Foraging' later in this section).

Founding — the queen builds an embryo nest

It may take 2 or 3 weeks for the queen to regain her weight and possibly migrate. During this period, the queen's ovaries begin to develop and the nesting drive intensifies until the queen selects a nest site (Spradbery 1973).

In the spring, the overwintered queen finds a location for her 'embryo' nest. In 2017, the first Jersey embryo nest was reported on 18th April, and in 2018, it was on 23rd April, at which point the nests had 8–10 cells and the first eggs had been laid; in 2022, embryo nests were again found in mid April.

The foundress is curled around the petiole, providing warmth to the eggs. Here, the outer envelope is not yet complete. Photo by John de Carteret.

Yellow-legged hornet nests are built, like those of other wasps and hornets, out of a kind of papier-mâché, made from harvesting plant material and pulping it with saliva and water. The papier mâché is then shaped into the structure of the nest,

*Left: View looking up into an embryo nest. The queen is again curled around the petiole. **Right**: Embryo nest attached to shed roof.*
Both photos by John de Carteret.

using the mouthparts. Because the plant material is from diverse sources, the colour of the material varies and results in coloured lines (browns, creams) that show how the material was laid down in thin strips, extending the edge of the paper. Crespo et al. (2022) looked at the nest material in detail and found that it was made almost entirely of lignocellulose, with beige material originating from softwood cells, while the brown strips were composed of hardwood cells, leaf tissues and grasses. The overall effect is usually a pale brown or beige-looking nest. Because water is so important for nest building, and the fact that the nest is constantly being expanded, yellow-legged hornets will often choose sites with good access to water in drier climates (Monceau et al. 2012). If there is water everywhere, this factor becomes less important (for example in Galicia, Rojas-Nossa et al. 2018).

The nest is built in a sheltered place as it is small, fragile and initially just big enough to house her first batch of workers. The embryo nest, built solely by the foundress, is between 3 and 10 cm in height, with one or two combs, and she adds around a cell a day to the structure. Cells in embryo nests and primary nests are smaller than those found in secondary nests (Dieguéz-Antón, 2022a). Embryo nests are usually built at a height below 10 m, perhaps reducing the energy requirements of the queen (it takes a lot of effort to fly up to a high nest carrying pulp or prey, Rome et al. 2015). Humans build all sorts of structures which create ideal shelter for this first nest, including barns, sheds, garages and eaves. Foundresses also like enclosed spaces like holes in walls or trees, and bird boxes continue to be popular. In south-western France (Andernos-les-Bains), 30 of 39 primary nests (77%) were associated with such man-made structures (Franklin et al. 2017). These are the places to keep an eye on in the spring. The foundress starts the nest by making a stalk or 'petiole' that extends down from a convenient beam, branch or ceiling. The cells are built at the bottom of this stalk and a thin, spherical envelope, at least two layers thick, is built around the whole lot to protect the eggs and larvae and keep them warm. The resulting nest is like a ball, with the entrance (of around 1.5 cm diameter) at the bottom (Diéguez-Antón et al., 2022a). Unlike honey bees, wasps and hornets build paper comb (the cells in which they raise their brood), which is horizontal rather than vertical, with the openings of the hexagonal cells pointing downwards.

The queen is alone and vulnerable until the first workers emerge, and during this queen colony or nest-founding phase, she hunts for food for herself (sap and nectar) and for the larvae (insect or other meat), lays eggs, and works on the nest. She will also spend hours resting, and curl herself around the petiole above the cells in order to provide warmth to the eggs and larvae. In another hornet, *Vespa simillima*, Makino & Yamane (1980) found that this behaviour could raise the temperature in the region of the cells by 2.5–4.0 °C. They suggest that such warming could accelerate the development of the immature workers, reducing the time the queen is alone and at risk in the embryo nest. Jersey data from 2022 show the first embryo nests being discovered around 3 weeks after the first queens were being spotted (Alastair Christie, pers. comm.).

Producing the first generation

Yellow-legged hornets, like many other insects, pass through different developmental stages before they become adults. The queen lays an egg, which she attaches to the bottom of the paper cell with a strong glue. Once the larva hatches from the egg it remains attached to the egg's shell, which prevents it from falling out of its downward-pointing cell. In later stages, the larva is prevented from dropping out by not wholly casting off its skin, which connects it with the egg shell. The larva will moult (cast off its skin in order to grow) four times, and is fed with balls of mashed-up insect meat, carrion and sugary liquids by workers, and grows rapidly. When the larva reaches its fifth stage, it is large enough to fill the entire cell, and it

Eggs in cells. Photo by Judy Collins.

Larva showing 'tail' made from previously shed skin.

Meconium. It is about 6 mm long x 5 mm wide x 4 mm tall. The paler material on the end is cast-off larval skin.

supports itself by pressing against the cell walls. At this point it cuts its attachment to the wall and is able to move freely within the cell.

Late-stage larvae regurgitate a special saliva when requested by the adults: this saliva contains important amino acids (the building blocks of proteins) and is a perfect food for the queen, workers and finally the sexuals (males and gynes) when they emerge (Matsuura & Yamane 1990). After completing feeding and prior to pupation, the mature larva applies thin silken lines around the walls of the cell, and spins a thicker cocoon cap over the opening. It dumps its gut contents into the bottom of the cell as it spins its cocoon and moults a final time. Thus, the larval excrement is sandwiched between the new cocoon and the final shed skin, forming a small dark brown package called a meconium, at the bottom of the cell. These meconia are not removed by workers when

Close-up of comb with older larvae and the silk-capped cells which contain pupae.

18

cleaning cells after adult emergence, and the queen will re-use the cell up to four times in secondary nests (Rome et al. 2015). The number of meconia found in a cell will therefore indicate how many times it has been used.

Inside its cocoon, the final-stage larva undergoes an incredible transformation, during which its whole body is rearranged into that of an adult. This is the pupal stage of development, and the pupa is at first cream-coloured like the larva, gradually becoming pigmented towards the end of its development. Eventually it bites its way out of the silken cocoon cap. The newly emerged adult (sometimes referred to as a 'teneral adult' or 'callow') is paler than older adults, has a softer skin and, in a similar hornet (*Vespa simillima*), spends the first 1 or 2 days head-down in an empty brood cell, resting and being fed by older

Adults biting their way out of their silk cocoons. Photo by Angus Deuchar.

An adult emerges from the cocoon. Its body hairs appear greyish. Photo by Angus Deuchar.

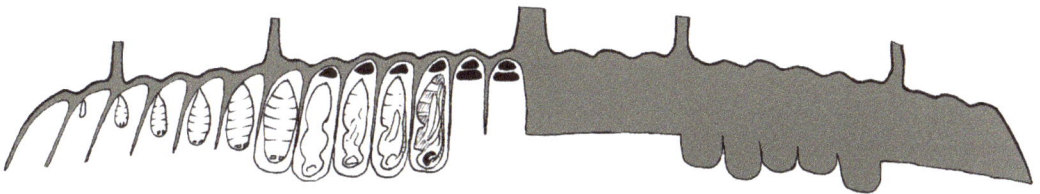

*Cross-section through a comb showing developmental stages from egg to larva, and finally pupa. The **black objects** in the bottoms of the cells are meconia. Notice that the fifth-stage larvu is free from its tether*

workers while its cuticle (skin) hardens (Martin 2017). Orientation flights in a variety of hornets occur most frequently on days 2–3 post emergence, once the wings have hardened. Immediately after, the hornets are ready to start contributing to colony duties (Matsuura & Yamane, 1990).

The development of the hornet from egg to adult takes around 50 days for the first cohort of workers (Thiéry et al. 2014, Martin in Prezoto et al. 2021), but as the nest grows, the weather becomes warmer, and the workers are able to control and maintain the temperature of the nest, this time becomes shorter. Dong & Wang (1989) found the different developmental stages in *Vespa velutina auraria,* to be 9–15 days for the egg, 10–18 days for the larval stage, and 15–20 days for the pupal stage in a laboratory caged primary nest (averages: eggs 13 days, larvae 15.8 days, pupae 19.3 days — overall 48.1 days from egg to adult emergence).

Early dangers for the young colony

The queen's embryo nest must be protected from the elements and small mammals, but the queen must keep leaving the nest to forage for food and nesting material. The queen is the sole provider, at first: if she dies, any larvae will die from starvation. Around Tours, in France, single queens founded 12 colonies, but out of these only three survived (Darrouzet et al. 2014).

Fragile embryo nest. The comb has come loose through handling, and is on its side.

One danger that might be faced by the young queen is that of nest usurpation — the taking over of the nest by another queen. Information from other *Vespa* species shows that if an embryo nest is destroyed, it may be that the foundress doesn't try to build a new one but instead tries to find another, and fights the resident foundress for possession. Such fights often cause injuries and death, and dead foundresses can be discovered beneath embryo nests (14 dead foundresses were recorded under an active *Vespa affinis* embryo nest, Martin 2017). Even the winning foundress may be injured sufficiently that the embryo nest fails (Martin 2017). Another possibility is that later emerging foundresses, while searching for a suitable nest site, come across an embryo nest and gamble on winning the fight and taking over the nest. Interestingly, Diéguez-Antón et al. (2022a) found two queens in four of the 62 embryo nests that they studied in Galicia, northern Spain. Were these queens competing or co-operating? They point out that two queens in a nest may offer a better survival strategy for small nests against predators: such 'polygyny' is a more common strategy in tropical hornets (Martin in Prezoto et al., 2021). Usurpation can happen up to a time when there are just a few workers. After that, the nest becomes too well-defended.

Very small primary nest with the first workers. Photo by John de Carteret

In the European hornet, queens emerge over a relatively long period — a 'bet-hedging' strategy because some of them will emerge from hibernation to better conditions than others (Gourbière & Menu 2009). It is not known whether the same is true for *Vespa velutina*. Foundresses appear and are seen foraging over a longer period than might be expected if they all emerged around the same time. On the other hand, you would expect them to be on the wing for quite a while, feeding, possibly migrating, setting up an embryo nest, raising the first cohort of workers and then flying for another 2 or 3 weeks after the workers have emerged — in total this adds up to perhaps 2.5 or even 3 months before the queen concentrates solely on egg-laying in the growing primary nest.

If the queen dies at the primary nest stage or later, and there are workers present, some of the workers may lay unfertilised (haploid) eggs that will develop into males. Early males have been commonly found in the European population (Monceau et al. 2013a), and when there is a queen present, these males are usually diploid and sterile, due to incomplete genetic material. For a little more detail about the ways that males and females are produced, please see 'Production of the sexual stages (fertile males and gynes)', p39.

Workers are hostile towards early males, attacking them (de la Hera et al. 2023) and seeming to drive them away from the nest, perhaps because males do not help with foraging, feeding, defence or nest building (Monceau et al. 2013b).

The primary nest grows

As soon as the first workers emerge, the whole process of nest building and brood rearing speeds up, and this speeding up carries on through the season, so that by the autumn, the development time from egg to adult reduces to around 29 days (Martin, in Prezoto et al., 2021).

This is partly due to more workers constructing cells and feeding larvae, and partly due to thermoregulation in the nest, as well as external temperatures getting progressively warmer. Like bees and wasps, hornets can produce heat to warm the nest by using their flight muscles. In another hornet (*Vespa simillima*), by the time the colony is producing sexuals, the temperature in the nest is being kept between 25 and 30 °C (Martin 1990). A warm nest not only speeds up development of larvae, it also means that individual hornets do not have to warm themselves up before they are ready for action, whether that is foraging at cooler temperatures during the day (unlike the European hornet, YLHs do not fly at night) or defending the nest at any time of the day or night. Larvae and pupae also appear to play a part in keeping the nest warm (Gibo et al. 1977).

If the nest gets too hot on summer days, the workers lower the temperature of the nest through evaporative cooling. To do this, they regurgitate water onto the outside walls of the nest (sometimes soaking them), and evaporate the water through fanning with their wings; they also ventilate the nest by fanning at the entrance (Perrard et al. 2009). This ability to regulate the temperature of the nest is one of the factors that make them such a flexible and formidable invasive species.

A primary nest in a building void. Photo by John de Carteret.

The queen will still leave the nest to forage once

Yellow-legged hornets on a primary nest: some are involved with nest construction (see darker band of papier-mâché which is still wet), while others patrol the surface. Photo by Francis ITHURBURU (CC BY-SA 3.0).

the first workers have emerged (behaviour that carries on for 2 or 3 weeks): after that she will stay in the nest (Martin 2017), except for moving to the secondary nest if one is built. The workers take on all nest duties, except for laying eggs: they forage for protein and sweet liquids, and for water and woody nest material; and they construct and remodel the nest, which is ongoing until late September (France: Monceau et al. 2017). They also guard the nest, adjust its temperature, clean the nest and one another, and share food.

Monceau et al. (2013b) found that older hornets became the first line of defence for the colony, being the first to react and come to the entrance when a caged colony was threatened (by a pair of tweezers!). Another study on a caged nest (Perrard et al. 2009) found that the majority of activities consisted of foraging (47%), nest construction (16%) and exploration (13%), while the remainder involved nest guarding, cleaning, ventilation, grooming and nestmate interactions. No outside activity was seen when the temperature dropped below 10 °C. In Galicia, foraging was seen at temperatures as low as 5 °C, and as high as 38 °C (Ana Diéguez-Antón, pers. comm.). Monceau et al. (2017) found that nest patrolling and nest building were pretty constant during the day, with slightly more happening during the middle of the day, but that much more effort was put into foraging. All activities increased over the summer and peaked in early autumn.

Although yellow-legged hornets do have castes (the workers are female, but perform different duties over their lifetime compared with the female queens) they do not seem to have the high level of age-related specialisation of tasks equivalent to that found in honey bee colonies (temporal polyethism). However, when studying the Oriental hornet, *Vespa orientalis*, Ishay & Ikan (1968) found that there were three types of behaviour that always occurred in the same sequence in the workers: (1) building or repairing cells; (2) caring for larvae; and (3) foraging and guard duty.

Recognition of nest mates appears to be based on complex cuticular hydrocarbons (CHCs). These chemicals are in the thin waxy layer that covers the cuticle (hard outer 'skin' of an insect). Such kinship recognition is probably important for these social insects where the workers labour on behalf of the whole colony (Couto et al. 2017). These and other forms of chemical communication within the nest are important for caste recognition and a variety of behaviours (Rodriguez-Flores et al., 2021). Interestingly, the variability of CHCs does not seem to have been affected by the genetic bottleneck that has occurred in the European population. This leads Gévar et al. (2017) to wonder whether the variability in CHCs arises from environmental effects such as diet and/or their gut microbiomes (gut microbes). Cini et al. (2020) found that different life stages and castes had distinct yeast and bacterial gut communities, and Zhang et al. (2022) found that lactic acid bacteria were significantly enriched in the midgut of male *Vespa velutina nigrithorax*.

The nest continues to grow, with a combination of fresh material added constantly to the outside, and material removed from the inside to make more comb. Like embryo nests, primary nests have their entrances at the bottom.

Relocation

If the space that the primary nest occupies is big enough, and the location is deemed secure enough, the nest simply expands right through the season, and the initial embryo nest built by the foundress becomes part of the larger secondary nest (Rome et al. 2015). In this case, the entrance hole migrates to the side of the nest at some point. However, the majority of primary nests are abandoned, and a new secondary nest is built. Dissection of

Bob Hogge holds a secondary nest, which has engulfed small oak branches and twigs. This one was around 45 cm in diameter. The entrance hole can be seen in the centre. Photo by Gerry Stuart.

69 nests in France, and the INPN database in France, suggest that approximately 70% of colonies relocate in the summer (Rome et al. 2015). Relocated nests can be distinguished from primary nests that just got bigger because the first (top) comb has a very regular structure in relocated nests; otherwise the embryo comb becomes embedded in the nest structure (Rome et al. 2015).

In *Vespa crabro* and *Vespa simillima* in Japan, once the decision to move has been made, some workers stop their normal duties and start scouting to find a suitable location for the secondary nest. It is assumed that *Vespa velutina* follows this pattern. Good sites (usually within 30–100 m of the old nest: YLH in Jersey) are marked by scouts settling there, and there is some back-and-forth with workers and scouts visiting possible sites and the primary nest until some sort of consensus is arrived at. After 3 or 4 days of this process, the queen leaves the old nest and searches for this new nest site. When she finds it she does not return to the primary nest, and her arrival triggers the workers into constructing the secondary nest. Workers left behind in the primary nest also have to find the new site for themselves, but once they have, there is traffic between the two nests for up to a month (presumably longer in *Vespa velutina*) while all the eggs, larvae and pupae develop into workers and find their way to the new nest. If the new nest is taken away at night during this relocation period, workers from the primary nest will build a new nest at the site of the removed nest (Matsuura & Yamane, 1990). It seems that the European hornet, *Vespa crabro gribodoi* (the British colour form) also relocates its nests, starting in small cavities or small mammal burrows and building secondary nests in larger cavities (Pawlyszyn 1992). However, they also build similar embryo nests to *Vespa velutina* in sheltered locations such as sheds and barns: make sure you know which species you are dealing with.

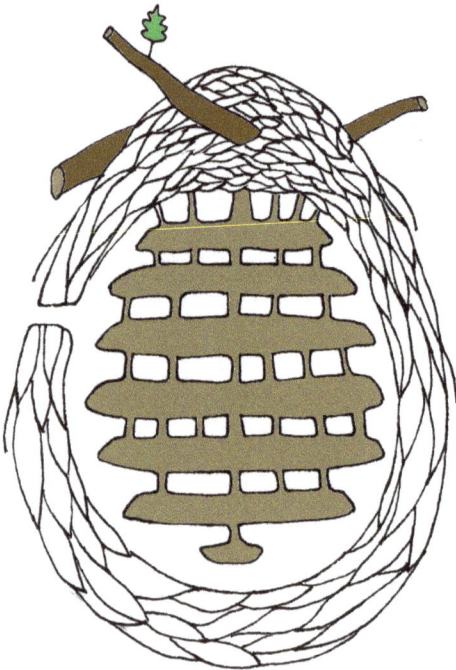

Cross-section through a secondary nest. Note the more compact paper pockets at the top, entrance hole on left-hand side and unfinished pockets.

26

The secondary nest

Secondary nests start as small balls that expand throughout the season, often becoming more elongated into an egg shape when they get big. In the early stages of their construction, they are characterised by a large number of workers, and the combs are without meconia, unlike primary nests (Diéguez-Antón et al., 2022a). In contrast to primary nests, or the nests of European hornets, the entrance is on the **side** of the structure in a secondary nest: it is usually circular and around 2.5 cm in diameter. According to observations from Jersey, the entrance holes often face roughly east, perhaps to facilitate early morning warming (Jones & Oldroyd, 2006), and tend to be below the meridian in tree locations, but above the meridian when the nest is located in brambles (Alastair Christie, pers. comm). Interestingly, Diéguez-Antón et al. (2022a), in Galicia, found some secondary passageways in addition to the main entrance, and these were in the bottom half of the nest.

Because the nest is in a constant state of construction, the outside has very distinctive structures that look like hanger openings. It is tempting to see these as entrances into the nest, but in fact they are half-finished 'bubbles' or pockets, which the hornets add to the outside of the nest walls to expand the nest. When complete, these trap air and insulate the nest. The nest walls at the sides and bottom are variable in thickness: around 1.5 to 5 cm in thickness (my obs. and Diéguez-Antón et al., 2022a). The top of the nest is made up of a dense arrangement of lots of small paper pockets, resulting in a thick, strong roof, (several times thicker than the walls lower down) while the rest of the nest has lighter walls made from fewer, but larger, paper pockets. Still, the whole nest is strong — more like light cardboard than paper. This material does not seem particularly waterproof (indeed, when they cool the nest with water, it sinks into the paper), but the exposed sites they choose in the tops of trees attest to the strength of the nests, even when wet. The complex wall structure must be strong enough to keep the nest rigid until it dries out in the wind.

Internally, the paper material of the nest may also help to buffer temperatures by absorption and evaporation of moisture in the air inside the nest [Klingner et al. (2005), for *Vespa crabro*].

The first horizontal comb is built on a single petiole and subsequent combs are built below: the youngest is at the bottom. These further combs are attached to the comb above by multiple paper columns. These are extremely tough pillars that are reinforced with old silk cappings from used brood cells (Martin 2017). As further combs are built below the first, additional struts are added between the top

Secondary nest cut away to reveal combs inside.
Photo by Judy Collins.

comb and the roof to carry the extra weight of combs full of brood. The brood is arranged in concentric circles, with pupal stages covered by their distinctive domed silk caps of varying heights.

Mature, late-season secondary nests are usually very big and distinctive, once you spot them, but secondary nests do vary in size enormously. One particularly large nest found in Aquitaine, France was 1 m tall and 80 cm across (Rome et al. 2015), while one measuring 1.2 m in height with a perimeter of 1.62 m was found in Galicia (Diéguez-Antón et al., 2022a), and a typical secondary nest from a tall tree and dissected in Jersey in 2018 was approximately 48 cm tall by 42 cm wide. Interestingly, the nests found in the UK in 2018 were all fairly small: 20–25 cm in diameter (Nigel Semmence, pers. comm.), but full-sized ones were found in 2023. Although secondary nests are stunning structures to behold, they are still very difficult to pick out amongst foliage, especially high up in trees.

Nests can be found in forests and agricultural areas, but in France 48.5% were discovered in urbanised habitats (Rome et al. 2015). It may be that nests are more likely to be spotted when more people are around to spot them, but a higher number of nests do seem to be occurring in urban and semi-urban areas (e.g. around villages) in France. Martin (2017) cites research that he was part of (Choi et al. 2012), which looked at the dominance of yellow-legged hornets among five species of hornet in South Korea. In this case, YLHs became more dominant as the habitat became more urbanised, with YLHs being almost the only type of hornet found in cities. It is not known, however, whether YLHs prefer urban environments in South Korea, or whether they are taking up residence in cities to avoid competition with other species of hornet in non-urban areas.

A study by Monceau & Thiéry (2017) looked at distribution of nests in the small coastal town of Andernos-les-Bains, west of Bordeaux, France. The study looked at data over 8 years (2007 to 2014), with a total of 528 nests. They found that in six of the years, the nests were randomly distributed within the town, suggesting that competition for nesting sites and/or food supplies did not exist. When the data were combined for all 8 years, they showed an aggregation of nests close to the seafront where there was an oyster farm and sport

A comb from the middle of a fairly large nest removed late September (Jersey). The brood is arranged in concentric circles with the oldest pupae at the inner sides of the rings of pupal domes.

fishing activities that favoured colony development. By the end of the study, the density of nests had reached 12.26 nests per square kilometre. The same data were used by another group to try to estimate the carrying capacity (i.e. the number of nests that an area can support, long term), and they arrived at an average density of 10.64 nests per square kilometre if nests were not detected and destroyed and 8.07 nests per square kilometre if they were. When this was adjusted to the purely *urban* area, a carrying capacity of around 23 nests per square kilometre (with ongoing nest destruction) could be expected. Interestingly, they also calculated that the relatively high detection rate for nests was only detecting about half of the nests actually present, and only 37.4% of active nests were discovered (Franklin et al. 2017: nests were not actively searched for). Bertolino et al. (2016) found a density of 2.9–3.5 nests per square kilometre in Italy. Densities of up to 15 nests per square kilometre are now regularly being found in French urban areas.

Franklin et al. (2017) also looked at the structures that nests were built on. Out of 39 primary nests, roughly 77% were located in or on man-made structures, and the remaining 23% on natural structures. For secondary nests, these proportions reversed, so that (out of 225 secondary nests) roughly 78% were on natural structures (of which roughly 94% were trees and 6% were other natural structures, like bushes and hedges) and 22% were on or in man-made structures. Three nests (1.5%) were found underground. The trees that were most popular were oak (64.3%), pine (15.9%), plane (4%), and poplar (2.4%): other species were represented at smaller percentages.

In Galicia, where yellow-legged hornets are now widespread, Diéguez-Antón et al. (2022a) determined that 59% of nests were found at altitudes below 200 m, 25% were found between 200 and 400 m, and 16% at altitudes above 400 m. The nest sites (all types of nests) were mostly located in vegetation (61%), and the rest in buildings (36%) and soil (3%). Most of the nests found in buildings were discovered in spring, coinciding with the development of embryo nests and included sites such as manhole covers, roof areas, unused attics, walls, waste containers, chimneys, windows, balconies, and even beehives. On the other hand, the nests found in autumn and winter were located mostly in deciduous vegetation or evergreen trees and shrubs. A nest in Jersey was found on a sea cliff around 4 m above the high tide mark.

Where yellow-legged hornets have been established for a few years (in France and Galicia), people have discovered rare 'hotspots' or favourite nesting sites where they will build nests in the same tree year after year, or even in the same year.

Hotspot. This tree in France had four nests in it, which were all active during the same season. Photo by Felix Gil.

By adding up the estimated numbers of eggs, larvae and meconia for each comb, 21 mature nests (meaning they had already started to produce sexual individuals) taken down in France in October were found to have an average of 6.3 combs, around 5,000 cells, and were estimated to have produced an average of around 5,300 individuals up to that point. However, the sizes of the nests did vary widely, resulting in a wide range in the estimates of individuals (minimum estimated as 82, and the maximum 13,340, Rome et al. 2015).
In Galicia, the largest nest found had 21,883 cells (calculated), and a whopping estimated population of almost 25,000 individuals for the whole season (Diéguez-Antón et al. 2022a).

The size of the nest and the population depend on how early the queen gets started in the spring, i.e. the larger nests will be in areas that are more suitable for the queen to start her embryo nest earlier, and the smaller nests will be in areas where the climatic characteristics are more adverse. In the case of Galicia, the largest nests were in coastal areas at lower altitudes and with milder climatic characteristics, coinciding with a longer colony cycle length, which gives the hornets more time to create much larger nests, with the queens beginning to emerge in February and the last workers dying in the January of the following year (Diéguez-Antón, pers. comm.). Farther north in Europe, hornets are more likely to be active for 9–10 months.

In a French study, the number of workers reached a maximum between late October and early November, with an average of around 436 workers present (Rome et al. 2015). A fairly large nest (around 50 cm tall and 53 cm across) removed from a tree in Jersey on 20th September 2023 contained 2,059 workers as well as the mother queen, 221 males and six gynes. Undoubtedly this is not a maximum possible number of workers: larger nests will have more. In the Alresford nest in the UK, which was removed 22nd September, 122 adults were found (the nest was 25 cm in diameter).

It is not known exactly how long *Vespa velutina nigrithorax* workers live for, but Dong & Wang (1989) give 24–142 days for life spans of workers and 4–60 days for life spans of males, all from a captive colony of *Vespa velutina auraria*.

Foraging

Yellow-legged hornets are generalist foragers, with adults temporarily specialising on certain meats (insects, carrion) and sources of sugar (sap, nectar). They explore widely and use visual and olfactory (smell) cues to return to profitable patches of food and detect prey (Wang et al. 2014). However, they don't seem to recruit others from the nest to exploit patches of food and, so far, nothing like the sophisticated waggle dance of honey bees has been discovered. Instead, they seem to operate individually, perhaps using some odour cues from returning foragers but this matter is not completely resolved (Richter 2000).

The distances they travel have been investigated intensely because foraging has such a bearing on honey bee predation and control methods, including finding the nest. Evidence for the distances they cover comes from several sources. Plotting observed individuals on a map around a nest is only possible when you know you have just one nest and they come from it. Individuals can be marked and timed to the nest and back (Jersey), or individuals from a captive nest can be fitted with tiny electronic RFID tags (Poidatz et al. 2018a). In this case, the hornet walks through a tunnel when leaving or entering the nest, where a reader can identify each hornet. This allows foraging trips to be timed, or hornets can be released elsewhere and timed home. A different approach is to use a flight mill to find out how fast they can fly, and for how long. A flight mill consists of a small stand with a freely rotating wire arm to which the insect is attached by the top of the thorax: without a surface for the legs to cling to, they generally start to fly by themselves, spinning around the stand on the wire they are attached to. The number of spins is converted to distance.

A consensus is forming that the majority of foraging occurs under 700 m from a secondary nest (Budge et al. 2017, Sauvard et al. 2018, Poidatz et al. 2018a), and perhaps 300 m from a primary nest (early summer, Sauvard 2018). Budge et al. (2017) found 60% of hornets were foraging within 500 m of the nest, and 98% within 700 m. Lioy et al. (2021a) used a harmonic radar to obtain a mean foraging range of 395 ± 208 m, with a maximum value of 786 m. However, there are also long-range foragers that will go up to 5 km (Poidatz et al. 2018a).

The same RFID study also found that only approximately 5% could get home if released 5 km away, and their return did not seem to be influenced by the direction of the release point from the nest (NW, NE, SW, SE), perhaps showing that they use visual and scent landmarks rather than sun and geomagnetic cues. They also found that the ones to make it back to the nest tended to be smaller and therefore were probably older and more experienced. Around 50% or fewer were able to get back to the nest when released at or more than 1 km away.

When they looked at times spent foraging, they found that the vast majority (95%) of trips from the nest lasted less than an hour, and that there was a tiny amount of nocturnal activity (2%: but this may have been a walk rather than a flight). Finally, there were hornets that went for particularly long trips, perhaps scouting, and some which did not come back for days — did they rest for the night in vegetation? And what of the hornets that did not return during the experiment: did they die or live as solitaries, or were they accepted by another nest? Perrard et al. (2009) found that some YLHs introduced from other nests to a caged nest were accepted.

Sauvard et al.'s (2018) study of tethered hornets on flight mills gave an average flight speed of 1.56 ± 0.29 m per second, but the authors note that in field observations, *Vespa velutina* is more likely to fly at speeds similar to free-flying *Vespa crabro*: around 6 m per second. This illustrates how different flight mill experiments are from real life. Lioy et al. (2021) used harmonic radar to measure flight speeds of yellow-legged hornets. This must be the most accurate way of measuring near-natural flight so far available. Several hornets were 'tagged' (had a small metal walking-stick-shaped device attached to the top of the thorax). This weighed only 15 mg — about a quarter of the weight of a normal prey-ball that is carried back to the nest. Yellow-legged hornets can generally still fly well carrying 80% of their body weight (Kennedy et al., 2018). The researchers' results found an outward-bound foraging speed of 6.66 ± 2.31 m per second, and a return speed (presumably with a prey pellet to take back to the nest) of 4.06 ± 1.34 m per second.

Yellow-legged hornet investigating a banana flower on Jersey. Photo by Gerry Stuart.

Foraging activity is throughout the day, but does seem to have a different pattern when temperatures get hot. A typical example of foraging hours in France is given by Monceau et al. (2017): 8:00 am to 8:00 pm, with a small peak in all activities, including foraging, between 1:00 pm and 2:00 pm.

In Galicia, Spain, hornets were seen all day in an apiary (from 6 am to around 9 pm) in the relatively cooler months of September and October, but during the hotter months of July and August there was a clear reduction in numbers of hornets seen during the hottest part of the day (for example, between noon and around 6 pm in July) (Diéguez-Antón et al. 2022b).

Food for adults

In order to sustain the rapid growth of the colony from the summer to the autumn, the workers forage for sugar-rich liquids for themselves and protein for the larvae. They are able to exchange food with other adults mouth-to-mouth in the nest (a process known as trophallaxis) and obtain food in the form of special saliva from later-stage larvae. Jeong et al. (2020) analysed this saliva in *Vespa velutina*, and found that it contained a high proportion of proline, used for endurance when hunting all day, and branched-chain amino acids that are directly metabolised in muscle as an energy source.

Feeding behaviour is tightly bound between adults and larvae. One could say that the larvae act as food storage for the whole colony (Matsuura & Yamane 1990). Ishay & Ikan (1968) found that, in *Vespa orientalis*, colonies could not exist without the larval salivary secretion because adults do not possess the proteases (enzymes) necessary for protein breakdown, and must get the raw materials needed for nitrogen metabolism and egg production (in the case of the queen) from the larvae. When larvae were removed from a colony, the workers left and the queen starved to death. Perhaps the foundress is furnished with enough amino acids to see her through the first stage of egg laying, before the first larvae are mature enough to produce the special saliva, or perhaps she obtains enough from nectar sources (Bouchebti et al. 2022).

Poidatz et al. (2022) used rubidium and caesium to label proteins and sugars, respectively, and so were able to follow the fate of these two types of food in the nest. The larvae, as expected, took in more protein than sugars, while the workers took in more sugars than proteins. Interestingly, the lightest larvae received more protein than the others, while the lightest workers received more sugars and the heaviest workers received more protein.

One external source of sugars that adults rely on is tree sap. When gathering nest material, fresh plant material is often moist with sap, and trees exude sap when broken, or wounded by birds, insects, mammals or fungi. Floral nectar is another source of high-energy food. This is limited to those flowers where the nectaries can be reached by the hornets' short tongues: *Camellia sasanqua* and ivy were the most obvious in Jersey (banana flowers were attractive, although they appear to have deep flowers). They also seek out human foodstuffs like sugary pop drinks and alcoholic drinks in pub gardens. A list of flowers used by yellow-legged hornets in Japan (Tsushima Island) and South Korea can be found in Ueno (2015). In Spain and Portugal, *Camellia*, *Eucalyptus*, loquat, common ivy, fennel, *Citrus*, and *Prunus* are all visited (Diéguez-Antón 2023). Ripe fruits such as grapes, plums, figs, pears and apples are gnawed (Nave et al. 2022). Indeed, fallen apples are a likely place to find adults in the autumn if there is a nest nearby.

Food for larvae

Hornet larvae are primarily fed with chewed prey pellets, and signal to the adults when they are hungry by rasping on their cell walls with their mandibles, producing a noticeable rustling/scratching sound (Matsuura & Yamane 1990). In the Oriental hornet, *Vespa orientalis*, if larvae are too small to signal in this way, but are hungry, workers looking after them will tap on the side of the cell with the tip of their abdomen to signal on their behalf (Ishay & Ikan 1968). These authors also looked at how often *Vespa orientalis* larvae were fed, and found that small larvae (first and second instars) were visited on average 53 times an hour; medium-sized larvae (instars 3 and 4) were visited 74 times an hour; and fifth-instar larvae were visited about 98 times an hour.

Yellow-legged hornet feeding on ivy flower. Photo by Juliette Poidatz.

As mentioned, early workers are small, resulting from less food being available. As the season progresses, more workers forage, so the larvae, although more numerous, get more food, generally resulting in bigger adults in the late summer and autumn. In fact, their mean weight can double (188.8 to 386.4 mg) from June to November (there is a lot of variability in these weights: Rome et al. 2015).

There have been plenty of observations of yellow-legged hornets taking meat from wild carcasses and from butchers and fishmongers in outdoor markets. In urban areas there are many opportunities for scavenging other sources of discarded protein too. But when it comes to hunting, yellow-legged hornets are formidable, and this is what causes concern for both beekeepers and conservationists.

Prey may be caught on the wing or on a surface, and is grabbed and killed by biting. After capturing its prey, the hornet worker either takes the whole insect back to the nest or processes it to some extent, from removing some parts to making a flesh pellet exclusively from the protein-rich thorax (which contains the big flight muscles). To make a 'meat ball', the hornet retreats to a leaf or twig and hangs by one or two back legs, using its other legs to manipulate the body of its prey. It removes some or all of the following: the legs, wings, abdomen, head and sometimes the thorax cuticle (the hard outer shell of the thorax). When removing all of the above, the process takes around four and a half minutes for a honey bee caught by a hornet in a captive nest (Perrard et al. 2009).

What are larvae fed?

Villemant et al. (2011b) found that the yellow-legged hornet consumes a great variety of insects, with an overall preference for social Hymenoptera (honey bees 37%, social wasps 18%) and flies (34%, mainly hover-flies, blow-flies and house-flies and their close relatives). Predation on honey bees was worse in an urban environment where there was less overall choice of prey. Some studies have caught and analysed pellets brought back to the nest by returning foragers: Villemant et al. (2011b) collected more than 2,500 prey balls between 2007 and 2009 in **urban**, **agricultural** and **forested** areas in France, with different results, depending on the environment (see pie charts below).

Perrard et al. (2009) collected loads from hornets returning to a wild **urban** nest during July 2007. They collected 235 loads, of which 71.8% were prey and 27.2% were pulp for nest building. Of the prey, 84.8% were honey bees, 11.7% were other insects (mostly sweat bees and flies) and 3.5% were vertebrate flesh, of which a third contained feather remains.

In a move towards a different technique, Rome et al. (2021) studied 16 nests in France between 2008 and 2010, identifying 2,151 prey pellets by a mixture of morphology (using pinned specimens and descriptive keys to work out what the prey was) and metabarcoding. This last technique uses DNA extracted from samples and matched with a library of samples in a large online database. They found that 38.1% consisted of honey bees, 29.9% consisted of flies and 19.7% consisted of social wasps, along with a wide spectrum of other animal organisms, mainly insects and other arthropods, but also vertebrate prey (presumably carrion). In total, 159 different prey species were identified.

Another two studies have used metabarcoding to find out what larvae are fed. Verdasca et al. (2021), in Portugal, extracted DNA from worker jaws (mandibles) and stomachs, and also from meconia (the larval gut contents that is excreted before pupation begins, and that is left in the bottom of the cell). Jaws were more reliable for detecting honey bee DNA than stomachs, and much easier to get from a foraging worker than the meconia that had to be obtained from a nest full of hornets a long way up a tree! The meconia showed DNA

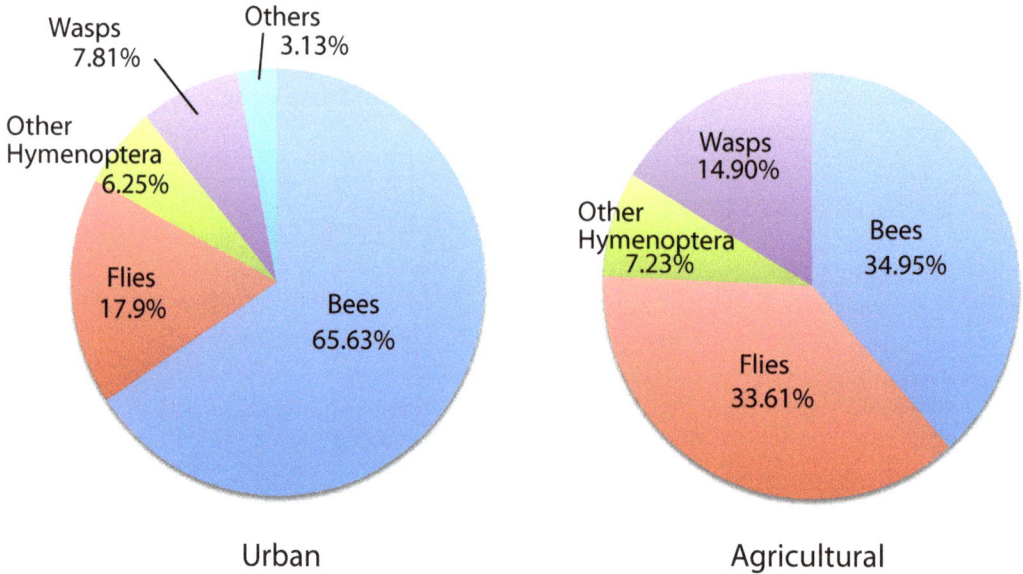

Wasps
7.81%

Others
3.13%

Other
Hymenoptera
6.25%

Flies
17.9%

Bees
65.63%

Urban

Wasps
14.90%

Other
Hymenoptera
7.23%

Bees
34.95%

Flies
33.61%

Agricultural

from all the prey that the larva had consumed before pupating, with honey bees being the most represented (all samples had honey bee DNA), followed by flies. Grasshoppers and moths were also present, with the latter likely to have been caterpillars when collected, rather than adult moths. Indeed, the authors point out that the ability of metabarcoding to detect soft-bodied prey items will probably make it less biased than the early attempts at identifying prey in captured prey pellets (e.g. Villemant et al. 2011b, or Perrard et al. 2009). The DNA does tell you that the species was consumed, but it doesn't tell you whether the prey was an adult or larva, or even whether the DNA came from an organism that was eaten by the prey the hornet consumed! With YLH metabarcoding, you tend to get either a snapshot of DNA on or in an adult, or everything that a larva has consumed since it hatched from the egg. Studies that collect pellets of prey intercepted from returning foragers can also give you information about amounts of different prey being brought back per unit of time.

Finally, a study by Stainton et al. (2023) used metabarcoding to identify prey by DNA found in the guts of larvae collected from five nests in the British Isles: three from southern England and two from the Channel Islands. They found 15–20 species of prey represented per nest, with wasps being the most abundant (and found in all nests), followed by blowflies and hover-flies, other flies (including tachinids) and spiders. There were some other types of insects, too, and hedgehog DNA, presumably from a carcass.

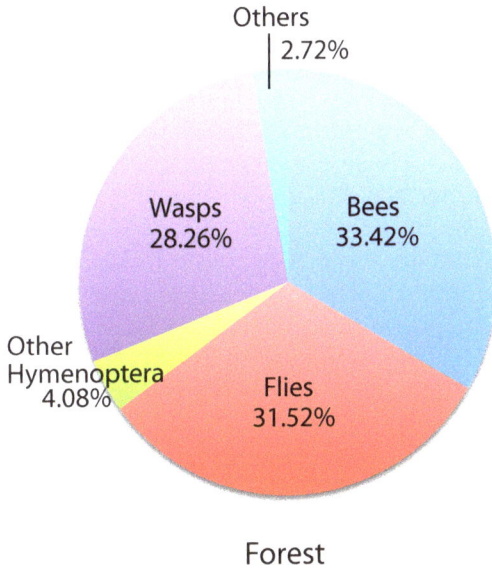

Forest

This page and opposite: pie charts after Villemant et al. (2011b). Note that 'Bees' includes wild bees as well as honey bees, but the wild bees (mainly bumblebees) are very few compared with honey bees (Claire Villemant, pers. comm.).

In a fascinating article, Cini et al. (2020) looked at the microbiome (in this case microorganisms found in the gut) of the yellow-legged hornet, and found that different life stages had different populations and types of fungi and bacterial groups, certainly affected to some extent by the very different diets that larvae and adults have.

Production of the sexual stages (fertile males and gynes)

Like other hymenopterans (bees, ants and wasps), hornets are haplodiploid. This means that their sex depends on whether they have two sets of chromosomes, or only one. Humans receive a set of chromosomes from their mother and a set from their father (whether you are biologically male or female), but it works very differently for hornets. If a hornet receives a set of chromosomes from its mother, and a set from its father (i.e., the egg has been fertilised: diploid), then it will be female, and will become either a worker or a queen. If the hornet develops from an unfertilised egg (i.e., it only gets genetic material from the mother), then it will be male (haploid).

This has consequences because workers can and do sometimes lay eggs if the queen dies — but these eggs always develop into males because the workers have not mated and so lack the sperm necessary to fertilise their eggs.

But there is a further twist in the haplodiploid story: the production of diploid males. If there is a loss of genetic diversity compared with the original population (as you might expect if a single queen founded the entire European population), then there can be problems at the sex-determining locus, and you end up with what should have been a female being male instead. These diploid males are usually sterile and, even if not, may father sterile female offspring. Such diploid males might be expected in populations that have arisen from a genetic bottleneck, and indeed many have been found in YLH nests in France (Darrouzet et al., 2015) and England (Jones et al., 2020).

Darrouzet et al. (2015) collected nests in France and found that 15 of 22 nests (68%) collected between April and August contained early diploid males ('early' because you would not normally expect males to be produced before September). This can have detrimental effects on the colony because the males use a great deal of resources as they are growing up — resources that could have been put into the production of workers and therefore would have ultimately gone towards production of fertile sexuals in the autumn. And, as adults, the males do not contribute to work done in the nest, or foraging.

There is an interesting hypothesis that if normal, fertile males were to be produced early in the year (and not sterile diploid males as often found), then perhaps these males could mate with unfertilised queens (gynes) that have survived the winter, or perhaps with workers in colonies that have lost their queen (Darrouzet et al. 2015).

From the shiny, hairless state of some mother queens found in nests, it appears that *Vespa velutina* queens have a 'royal court' where workers attend the queen, licking and biting her, and even sometimes nibbling off her wings. Such behaviour seems linked to the production of large cells by workers, which are intended for gynes (Matsuura & Yamane 1990).

In mid to late August the queen starts to lay eggs that will become fertile males (i.e. normal unfertilised haploid eggs) and then she starts to lay fertilised eggs that will become gynes, (virgin queens that, once mated, will become the queens of colonies next spring). The mechanism that leads to the production of gynes rather than workers (they are both female) is not known, but there are several factors that might be involved. The clearest of these so far is the amount of food that larvae receive. As the colony grows, the ratio of larvae to workers (L/W) drops, which means that each worker provisions fewer larvae as time goes on, and the larvae become better fed. When the L/W ratio drops to around or below 2, sexuals start to be produced, the switch presumably triggered by the level of nutrition (information for other hornets: Matsuura & Yamane 1990). In addition to the amount of food received, pheromones, oral

Colony ('mother') queens that have become shiny through removal of body hair by workers' attentions. The queen on the left has also had her wings nibbled off. Photos by Chris Isaacs.

secretions and perhaps vibrational signals from adults during feeding of larvae, may play a role in triggering the divergence from the developmental pathway of a worker to that of a gyne (Greene, 1999; Jeanne & Suryanarayanan, 2011).

Although larger cells are built in the later combs at the bottom of the nest, gyne production is not limited to these (pers. obs; Rome et al. 2015); the silk caps do give a lot of leeway in terms of available room for pupation.

On average, three times more males than gynes were produced from mid-September to the end of November in France (on average 350 gynes vs 900 males, Villemant et al. 2011b). Herrera et al. (2019) found that 57.9% of adults found in nests were male after mid September in Mallorca. Chauzat & Martin (2009) mention a positive correlation between the nest size and the number of queens raised and, certainly, the longer favourable conditions continue in the autumn, the greater the number of gynes that will reach maturity. Hence the race to find and destroy nests in the autumn before they have a chance to produce the sexuals.

The reproductive strategy of hornets is based on sexually mature males and gynes leaving the nest around the same time (Poidatz et al. 2017), and for this to coincide with production of sexuals by nearby nests. Matsuura & Yamane (1990) report that males and queens of other vespine wasps often leave the nest at around the same time.

In France, the main sexual cohort of males emerged from their cocoons in early September onwards, about 15 days before reproductive females (Rome et al., 2015). During a male's first 10 days as an adult, the migration of sperm from testes to seminal vesicles is completed, and the males are then sexually mature. When this process is complete, they will leave the nest and start to forage for themselves (Poidatz et al. 2017). Newly emerged males and gynes will spend their first 1-2 weeks based in the nest, feeding on larval and worker regurgitations. The gynes, especially, need to build up their fat reserves to see them through their hibernation.

In the French study, both males and gynes reached maximum numbers during the first half of November and most had left the nest by the end of November. In December the last workers were still present and feeding the sexual brood, although due to prey scarcity, the brood was unlikely to reach maturity at this point (Rome et al. 2015).

After the sexuals have started to develop, at some point the mother queen disappears or dies and the nest is characterised by an irregular brood pattern and the appearance of two eggs being laid per cell

Male yellow-legged hornet showing underside with distinctive pair of yellow spots near the tip of the tail. Photo by John de Carteret.

Different castes. From left to right: queen, worker, male. Photo by Chris Isaacs.

(Diéguez-Antón et al., 2022a). The last nest that still had a mother queen in it was collected on 27th October in the French study (Rome et al., 2015).

Mating

Genetic analysis of nests removed in the UK in 2018 found that queens had mated with one or two males (Nigel Semmence pers. comm.), as you might expect when nests are scarce. In France, where nests are plentiful, gynes mate with between one and eight males (Arca et al. 2015). This behaviour of a queen mating with multiple males (a system called 'polyandry'), is very rare amongst hornets — *Vespa velutina* is almost unique in practicing polyandry, and it gives this species a major advantage when invading new territories where genetic 'bottlenecks' (severe restriction of available genetic versions due to too few unrelated individuals) are likely to occur (Arca et al. 2015). Indeed, *Vespa velutina*, along with a few wasp species (*Vespula vulgaris, Vespula germanica* and *Vespula pensylvanica*) all have an average mating frequency of 2 or above, and all three species have become major pests outside their native range (Martin in Prezoto, 2021).

In China, Wen et al. (2017) found two compounds that act as male-attracting sex pheromones in gynes of *Vespa velutina auraria*. They did this by using a combination of observations of 'feral' colonies and obtaining and analysing the compounds from newly emerged captive gynes. They found that gynes would leave their nests on sunny days and head for open areas at the edge of a nearby forest. The gynes would repeatedly fly out into these areas to attract males, then fly back to a more sheltered location within the trees. The gyne would rub herself with sex pheromone, which would attract swarms of males. After mating, she would no longer produce these compounds.

In northern Spain, researchers who were investigating interactions of *Vespa velutina* with ivy, and other insect visitors to ivy in the autumn, noticed mating occurring (Rojas-Nossa et al. 2023). The ivy seemed to be acting as a congregating area, attracting both males and gynes. A male would grab a gyne in the air and they would immediately drop to the ground where copulation would occur. The two hornets form an 'S' shape together while mating; in other hornets the two

Gyne and male mating on a window pane. The gyne is much larger than the male. Photo by Sylvie Richart-Cervera / INRAE SAVE.

participants take it in turns to stay still on the ground, or support their arched body with their back legs.

Males do not die as a result of mating, as they do in honey bees, and work by Poidatz et al. (2017) suggests that each male only mates once, and eventually dies around December. Cappa et al. (2019) found that males were attracted to bigger females with more abundant fat storage, regardless of how closely related they were. This was a surprising result and could help to explain the low genetic diversity and high levels of diploid males found in European colonies of *Vespa velutina*, although gynes might actively reject male nestmates.

It would be interesting to know whether some males disperse widely. It might make sense for some males to travel away from their nestmate gynes in search of gynes from other nests. There seem to be enough males to allow this. In England, in September 2018, a single male was found at least 12 miles from the nest that it seemed to have come from (using genetic analysis); in October, two males were found in Dungeness, suspected of being blow-ins (no local nest was found).

Nest decline

Once the sexuals have left the nest, the numbers of workers decline, due to a combination of fewer sources of food, colder temperatures, their own life spans coming to an end and lack of replacement by new workers, until the nest is fully abandoned.

Rarely, one or more gynes can be found in an empty nest; these have emerged too late to mate and often have atrophied wings. Since they are unmated, they will not be able to start their own colonies (Villemant et al. 2011b). The last of the larvae in the nest die of starvation as the food resources run out. When the numbers of foragers reduces, and workers in the nest have nothing to feed to larvae, sometimes larvae are ejected (Diéguez-Antón et al. 2022a) and can be found on the ground below the nest.

However, in Galicia, NW Spain, Diéguez-Antón et al. (2022a) and Sanchez & Charles (2019) investigated nests taken down in January that contained all life stages (eggs, larvae, pupae and adults), and all castes (workers, gynes and males). This extended cycle was also found to be true in Portugal (Nave et al. 2022). It is not known whether mated queens hibernate at all under these circumstances. Sanchez & Charles suggest that winter nests should be destroyed in this region because they may well still contain gynes capable of founding colonies, whereas farther north, in France, nests are fully abandoned by this time of year, and the gynes have mated and are hibernating.

The nests are not re-used (although there has been a case of an embryo nest being built inside a secondary nest which had been stored in an out building). Over the winter the nests become silvery-grey with age. With winter weather, many break up and fall to the ground. Green woodpeckers, jays and tits will eat the last of the larvae and pupae at the end of the season (Mollet & De la Torre, 2006).

Queen hibernation

After mating, the queens spend time searching for hibernation sites. There are few data on the dispersal range of queens. In Spain, Caragata & Montesinos (2020) trapped emerging queens in the spring, and caught the greatest numbers in the first 250 m from an old nest that wasn't destroyed the previous season, with no more caught at distances 500 m and greater from the old nest. Hibernating close to the old nest would conserve resources for hibernation, as it is thought that they don't eat after copulation. However, humans can accidentally transport them great distances if they hibernate in cargo. The ovaries of hibernating queens remain in an immature condition, and the sperm they have stored in their bodies is conserved for the colonies they will build next season. The fate of unmated gynes is unknown, but if they survive the winter, they may live as solitaries the following year (there is evidence for this happening in *V. mandarinia*, Matsuura & Yamane 1990).

A queen hibernating in a hibernaculum she has excavated under bark. Photo by Dominique Soete.

Ecology

The yellow-legged hornet is an invasive alien species

According to the International Union for Conservation of Nature (IUCN, the people who publish the red list of threatened species):

> *Invasive alien species are animals, plants or other organisms that are introduced by humans, either intentionally or accidentally, into places outside of their natural range, negatively impacting native biodiversity, ecosystem services or human economy and well-being.*

And they are

> *...one of the biggest drivers of biodiversity loss and species extinctions.*

Why has the yellow-legged hornet been so successful in Europe?

Wide range of food

The yellow-legged hornet is a predator that kills a wide range of insects, but can also choose from other sources of meat such as mammal, bird, fish and crustacean carcasses, and meat prepared by humans. It also pursues sources of sugar such as sap and flowers where the nectar is reachable by its fairly short tongue. But the clever bit is how these resources are used in the nest. Protein is fed to larvae, but there is a constant circulation of proteins and sugary carbohydrates between workers and larvae. Larvae are fed relatively unprocessed meat, and workers receive saliva from larvae that contains sugars and the amino acids they need to make proteins. And there is exchange of food between workers too (Poidatz et al., 2022). In this way, from a diverse range of foodstuffs brought to the nest, hornets of all ages get the type and quantity of nutrition they need, with larvae acting as both production vats and storage tanks.

Rapid spread

The speed with which yellow-legged hornets have spread through Europe has been surprising. This has been accomplished by a combination of queen migration after hibernation, and accidental hitch hiking on human transport, resulting in speeds of 18 to 78 km per year across continental Europe.

The nest and social insect advantages

The design of the hornet nest adds to their success. In it, they are able to keep the temperature warm and stable for maximum speed of larval development, and even cool it if necessary, allowing them to nest at a range of temperatures. Nest sites can be natural or artificial. The lack of trees in an urban setting does not limit them, and they can live in forests, agricultural areas, suburbs, towns and cities. Even the wood used to make the nest does not have to come from a living tree. Water, however, is limiting as it is needed to create pulp for nest construction and for drinking.

As social insects, they act together to produce a single batch of sexuals at the end of the season, and the only indispensible individual in the colony is the queen. Thus they can survive losses of workers, and sacrifice some larvae in achieving this goal (at the end of the season, some larvae feed sexuals but may not become adults themselves).

Few predators or parasites

Invasive species may be suddenly freed from their old enemies. In the case of the yellow-legged hornet, this includes some ants that will attack embryo nests and, more importantly, other hornets. In time, you might expect new predators to take advantage of such a plentiful resource as nests full of hornets, but that might mean an awful lot of time unless they already have a taste for Hymenoptera, such as the honey buzzard, yellow-throated pine marten or bee-eater.

Great immune system

Cappa et al. (2022) looked at immune competence between the YLH and the European hornet. They found that workers and males of both species showed a similar, lower response when challenged by the bacterium *Escherichia coli*, probably because they don't face the challenge of hibernation. However, yellow-legged hornet reproductive females (both gynes and foundresses) were much better than European hornet gynes and foundresses in their ability to clear bacteria from their systems.

Not limited by genetic bottleneck

Considering that the whole European population appears to have originated from a single multiply-mated queen, there seem to have been few genetic problems. That the queen had mated several times is probably key here. One might expect inbreeding to lead to sterility but, overall, this doesn't seem to have been a problem for them, despite diploid males often being found. In fact, according to Garnas et al. (2016), such founder effects rarely limit fitness in invasive insects, and may even benefit populations by purging harmful genetic material. Multiple introductions, even from far-flung invasion fronts of the same original population, can increase genetic diversity. This may have occurred when the population arriving in Portugal met that of the main front working its way along the north coast of Spain.

Little competition among yellow-legged hornets

One might expect that 'pioneering' yellow-legged hornets moving into new territory might have different population dynamics to established populations, where there might be more competition between colonies for prey and nesting sites. When Monceau & Thiéry (2017) did their study in Andernos-les-Bains (south west France), they found that as the number of nests increased from four in 2007 to 111 in 2014, the nesting sites (apart from in one year) appeared randomly located, which would suggest that there were no territories and thus no competition between the colonies. This was corroborated by Franklin et al. (2017) who, using the same data, suggested that the high densities of nests pointed to a lack of competition for resources. They calculated densities of around eight nests per square kilometre with continued destruction of nests as the eventual carrying capacity. But, as mentioned previously, in a purely *urban* area, a carrying capacity of around 23 nests per square kilometre (with ongoing nest destruction) could be expected.

Extremely difficult to eradicate

So far, the only successful eradication of an established population (i.e. one that has produced the next generation or more) has happened in Mallorca, mainly through nest removal. This was only possible because the emerging population was isolated, with no other nearby populations poised to move in to fill the void. In Mallorca, like the UK, they remain permanently on guard against further incursions.

What are the ecological impacts of the yellow-legged hornet?

Effect on native insects

The yellow-legged hornet is a prolific predator of medium to large insects. Although yellow-legged hornets tend to consume more locally abundant prey than anything else, they are opportunists, and may prey on rare local species, too. Some insect families, for example flies, seem to have more diverse members preyed upon (102 different species of Diptera taken, compared with 14 different hymenopteran species, Rome et al. 2021). The study found 159 different species in prey pellets, but another measure of diversity suggested that over 400 different species were predated by YLHs in the study area.

In the UK, our insect communities are already under threat from habitat loss, agricultural intensification, etc. At densities of 10 to 20 nests in a square kilometre (perhaps 100,000 hornets in that space over the season), these predators could put local insects under great pressure. The biggest consumption of insects by the YLH occurs within a restricted period from July to October, which may save some species from being targets.

Do native insects have any defence against capture by yellow-legged hornets? In a recent paper, O'Shea-Wheller et al. (2023) investigated the impact of predation by YLHs on colonies of bumblebees (*Bombus terrestris*) in northern Spain. Although grabbed in the air by hornets, the bumblebees responded by stopping flying, which resulted in both insects dropping to the ground. When they hit the ground, usually the hornet's grip would be loosened, and the bumblebee could escape. If that didn't happen, the bumblebee would assume a defensive posture, on her back with her legs in the air and sting raised, forcing the hornet to abandon the chase. Such behaviour avoided predation by the hornet (although it doesn't always work). However, they found that bumblebee *colony* weights were reduced where YLH numbers were higher. This could be due to the hornets disrupting bumblebee foraging, or perhaps through competition for resources; but lower colony growth did not result in colony loss, or prevent the production of queens.

Apart from direct predation, YLHs might affect native insects in more subtle ways, such as by competing for resources, or through harassment, intimidation and stress. Several authors have looked at how YLHs might affect the European hornet. In Jersey, YLHs feeding on tree exudates would leave if a European hornet turned up: European hornets are significantly bigger and above YLHs in the 'hornet pecking order' (Kwon & Choi, 2020). In fact, some authors speculate that European hornets could slow the spread of yellow-legged hornets where there are enough of them (Carisio et al. 2022). Although European hornets do occasionally take honey bees at hives, it's not something that beekeepers generally worry about. Barrachi et al. (2010) noticed that the European hornet sometimes partially chews the honey bee abdomen to extract the honey crop before discarding it. It would be fascinating to know whether this also happens in YLHs: it could be advantageous to obtain both nectar and protein from the same prey item, but this could be offset by risks of pathogens picked up from the bee abdomen.

The similarity of European hornet and YLH diets and nesting sites would suggest that there might be some competition between them (Monceau et al. 2015). Not only have European hornets been observed predating more on honey bee colonies since the introduction of YLHs into France (Monceau et al. 2014b), but they have also been seen scavenging on dead bees left by the YLHs (Monceau et al. 2015). Perhaps it might even be advantageous for European hornets to live alongside YLHs and such 'synergistic predation' only increases the losses for honey bees. The better immune response of YLHs (Cappa et al. 2022), might favour them over European hornets. If bacterial challenge is an indicator of overall immune competence, then YLHs may also cope better with other pathogens and parasites than the European hornet.

Yellow-legged hornet in flight. You can clearly see the two pairs of wings. Photo by Xabi @desde.mi.desenfoque.

The authors point out that if control of YLHs is attempted using microbial agents, the European hornet may come off worse.

In Italy, Carisio et al. (2022) found that where YLHs were at low numbers, European hornets did well, with greater numbers even than in places where YLHs were not present. Where YLHs were abundant, it was more difficult to see any relationship between the two species. This points to a lack of competition between the two hornets. It may be that the niches they occupy are different enough to avoid such competition when there are few YLHs around (a niche is a total environment occupied by an organism, including food sources, predators, places to nest, climate etc).

Lioy et al. (2023) looked at niche overlap between three hornets found in Europe: the European hornet, the YLH and the Oriental hornet, *Vespa orientalis* (now moving north from the very south). They concluded that, although there was a certain degree of overlap, the European hornet is more suited to colder, drier environments than the YLH.

On Tsushima island, in Japan, *Vespa velutina* has been present since 2013. On the island there are several species of native hornets and researchers wondered whether interspecies mating could be occurring between *Vespa velutina* and any of the native hornets. The study, by Yamasaki et al. (2019), found that, out of four native hornets tested, one species, *Vespa simillima*, had indeed mated with *Vespa velutina*. In fact, 43% of *V. simillima* queens tested contained *V. velutina* sperm, and 28% had **only** *V. velutina* sperm. They concluded that this reproductive interference could pose a threat to *V. simillima*, and might also affect other hornets by way of a shift in their relationships (predation occurs between different hornets).

In terms of wider competition for prey, yellow-legged hornets might also compete with some wasps (social and solitary), robber flies, spiders and insectivorous birds.

Effect on pollination

When I wrote the first edition of this book, there wasn't much information about the impact of yellow-legged hornets on pollinators or pollination. Since then, there have been a couple of important papers.

The first was a study by Rojas-Nossa & Calviño-Cancela (2020), who looked at patches of a late-blooming flower, apple mint (*Mentha suaveolens*), in northwest Spain. Yellow-legged hornets did not feed on the flowers, but used the patches as hunting grounds, where they were frequent hunters of flower visitors (8% success rate in 25 hunting attempts), flying fairly constantly and usually catching their prey in the air. Visitation rates by honey bees, bumblebees, small bees and hover-flies were all significantly reduced when the yellow-legged hornet was present, but wasps, even though they also suffered attacks, increased their flower visits in those patches. As a result of disturbance of pollinators, the quantity of useful pollen on plant stigmatas decreased in patches where the hornet was hunting; a clear indication that the hornets were reducing pollination by native insects.

Next, consider a situation where yellow-legged hornets visit a patch of flowers to hunt for pollinators *and* find nectar for themselves. In this case, they can disrupt the existing ecological web in even more ways: predation, intimidation, competition for sugary foods, and disruption of pollination.

Rojas-Nossa et al. (2023) looked at YLHs on ivy (*Hedera hibernica*) in northwest Spain. For the hornets, the ivy acted as a hunting ground for insect prey, a foraging area for nectar (over 64% of their time was spent doing this), and a congregation area for males and gynes. Where yellow-legged hornets were abundant, visits to flowers and time spent at flowers by native pollinators were significantly reduced, and the YLHs were the most common nectar feeder on the ivy. But they weren't as good at pollination as the native pollinators, and the ivy suffered reduced numbers of seeds per fruit as a consequence. Overall, the hornets had a predation success rate of 12.5%. Hover-flies were the most targeted insects, but hornets had a success rate of only 3.4% when catching them in flight. Some wasps and flies were attacked less often, but the hornets had a success rate of 50% with these. Attacks on insects were generally more successful when the victim was perched on a flower. Insects also responded differently to attacks, with blowflies leaving the plant after an attack, while those in the house-fly family, tachinid flies and bees continued to forage. Some wasp species were strongly attacked, but two thirds carried on foraging. This last point seems in disagreement with Carisio et al. (2022), who found wasp numbers to be unaffected by yellow-legged hornets.

Disease reservoir

There have been quite a few studies into the viruses that are carried by yellow-legged hornets. Yang et al. (2019) found aphid lethal paralysis virus (ALPV) in *Vespa velutina* and in the Eastern honey bee, *Apis cerana*, and suggested that the hornet was acting as a reservoir and vector for this virus. Pollinators that share resources (such as flowers) with YLHs could therefore be more exposed to insect viruses, with even the possibility of enhanced virus transmission and increased virulence, as has happened with honey bees and *Varroa* mites carrying viruses. Gabín-García et al. (2021) found that *Vespa velutina* harbours most common hymenopteran pathogens, as well as several new parasites. The parasites it carries are similar to those carried by the

European hornet. Due to the large numbers of yellow-legged hornets possible in an areas with, say, 12 or more nests per square kilometre, the authors suggest that the YLH could severely affect pathogen dynamics and could have long-term harmful consequences on native insects.

Yellow-legged hornet looking for nectar. Photo by Xabi @desde.mi.desenfoque.

Effect of trapping

When looking at the effects of the yellow-legged hornet on ecology, I should mention the corollary of the invasion of Europe by the hornet: the subsequent widespread use of killing traps, to try to reduce numbers of hornets. Intensive use of drowning traps could have a major impact on native insects, perhaps even greater than the impact of the hornet itself (Rome et al. 2021).

Bioaccumulation

Because yellow-legged hornets predate on a wide range of insects, it might be expected that pesticides and other agricultural chemicals, picked up by their prey, might accumulate in them. To test this idea, Tison et al. (2023) analysed YLH workers from 24 different nests, finding 13 different pesticides, including 8 fungicides, 3 insecticides and 2 miticides. Hornets from a nest in a sub-urban situation (23% urban, 39% grassland, 33% forest and 5% water) were the most contaminated. Bioaccumulation is also an issue in places where hornets are collected as a human food (e.g. China and South Korea).

Any winners?

Ecologically, are there any winners from the spread of the yellow-legged hornet? Well, yes. Some (a very few) animals already include hornets in their diet and are able to deal with this potentially dangerous prey. The honey buzzard, *Pernis apivorus* feeds mainly on wasps but, in 2018, a piece of hornet nest was seen in a camera image of a nest and retrieved for analysis. It turned out to be from a YLH nest (Macià et al. 2019). Rebollo et al. (2023) looked at nine European honey buzzard nests in north-western Spain and found YLH nest remains in all the nests: chunks of hornet and wasp nests are taken to the nest so that their young can feed on the larvae. In fact, YLH was the second most abundant wasp species in their diet, and the most abundant in 2018 based on biomass. During the breeding season, each pair of honey buzzards attacked 34–61 colonies in 2018 and 15–28 in 2019. Nesting frequency rose from 60% before the hornet's appearance to 100% afterwards, and the density of breeding pairs was three times higher after the hornet's arrival. In fact, Rebollo reckons that the 700 breeding pairs of honey buzzards could have destroyed as many nests (24,500) as destroyed by the local authority in 2018 (Onofre et al., 2022). The honey buzzard sometimes breeds in England, but is mainly migratory.

In parts of Europe, the bee-eater (*Merops apiaster*, a rare visitor to the British Isles) has also been found to predate YLHs, but catches adults on the wing and then rubs them against a surface to get rid of the sting.

Onofre et al. (2022) also mention other birds that could predate yellow-legged hornets. There have been records of the royal shrike (*Lanius meridionalis*), red-backed shrike, jay and magpie all feeding on European hornets. They also mention mammals that already prey on wasps: badgers, moles, wood mice, house mice, pine martens, stoats and weasels, which may destroy nests and/or eat hibernating queens: these may adapt to yellow-legged hornets as prey, but only where nests are reachable, such as embryo and primary nests (pine martens are probably the only ones who could reach some nests up in trees).

In South Korea, the yellow-throated pine marten was found to have eaten yellow-legged hornets, mostly males and new queens attempting to mate (Kim & Choi, 2021).

For those birds and animals that do take yellow-legged hornets into their diet, bioaccumulation of pesticides from insects that the hornets have predated, or direct ingestion of poisons from treated nests, is always a danger.

A yellow-legged hornet flies amongst honey bees in front of a hive in France.
Photo by Sylvie Richart-Cervera / INRAE SAVE.

Yellow-Legged Hornets and Honey Bees

Catching and killing honey bees

When honey bees can be such a large component of the yellow-legged hornet diet, you can see why British beekeepers are so alarmed at their arrival. Especially after seeing what has happened in France over the last 20 years.

Honey bees are perfect for YLHs — they're big, with plenty of meat on them, and humans keep them in these boxes in open areas where they are easy to attack. Not only that, but usually there are several boxes in one place! Tens to hundreds of thousands of prey items in one place — this is fairly easy, fast, food.

Yellow-legged hornets are 'central place' foragers, meaning that they need to keep returning to the nest to feed larvae, and in a world of patchy food supplies, a hive or apiary is a very efficient way of providing the nest with food. Landmarks, both visual and olfactory (scents of hives and other scents in the environment) are used to navigate to hives and back to the nest (Richter, 2000).

There is some evidence that individual hornets might stick to food types, at least temporarily. The NBU, when tracking marked individuals in England, used both fish fillets and wasp attractant. They found that the same individuals would consistently return to either protein or sugars, perhaps indicating that a worthwhile patch of food should be exploited until it starts to dry up. Within a food-type (in this case honey bees), Thiéry et al. (2014) found that more than half of attacking YLHs returned to the hive they were previously caught at, to carry on predating. However, Monceau et al. (2014b) found most of their hornets (88%) were recaptured in front of different hives.

Couto et al. (2014) found that hornets captured in front of hives were especially attracted to the odours of pollen, honey and brood, even more than they were attracted to the smell of adult honey bees, and meat and fish were ignored. Hornets were also attracted to the odour of geraniol, which is a component of the Nasanov pheromone in honey bees. Geraniol gathers bees to a swarm or the hive entrance, and may indicate to hornets the presence of large numbers of honey bees. These observations suggest that YLHs mainly find their prey through scent.

Once a hive has been located, hornets will hover in front of it and attack mainly returning foragers, a behaviour known as 'hawking'. Returning foragers may be preferred because they can be 40% heavier when they return with nectar and pollen loads, which will affect their speed and agility. Because foragers are the oldest workers, wing damage may also affect their flight (Monceau et al. 2013c). However, hornets will also attack bees on the ground and on the hive itself, especially if a bee gets isolated from the crowd. Hornets sometimes hide under raised hives, darting out to capture bees returning from foraging (Diéguez-Antón et al. 2022b).

Predation pressure increases over the season. From the first hornets spotted in an apiary, the period of intense predation begins around 40 days later (July in both France and northern Spain: Monceau et al. 2013c, Diéguez-Antón et al. 2022b) and extends to the end of October or November (French data), with a maximum in mid September and a drastic decline at the end of the season. Diéguez-Antón et al. (2022b) took photos of the fronts of two hives throughout the season in Galicia, northern Spain, and noticed a few hornets appearing in photos a full month before they were noticed by the beekeeper.

Monceau et al. (2013c) found that in France, during the course of a day, hornet numbers in front of hives were fairly constant, but they

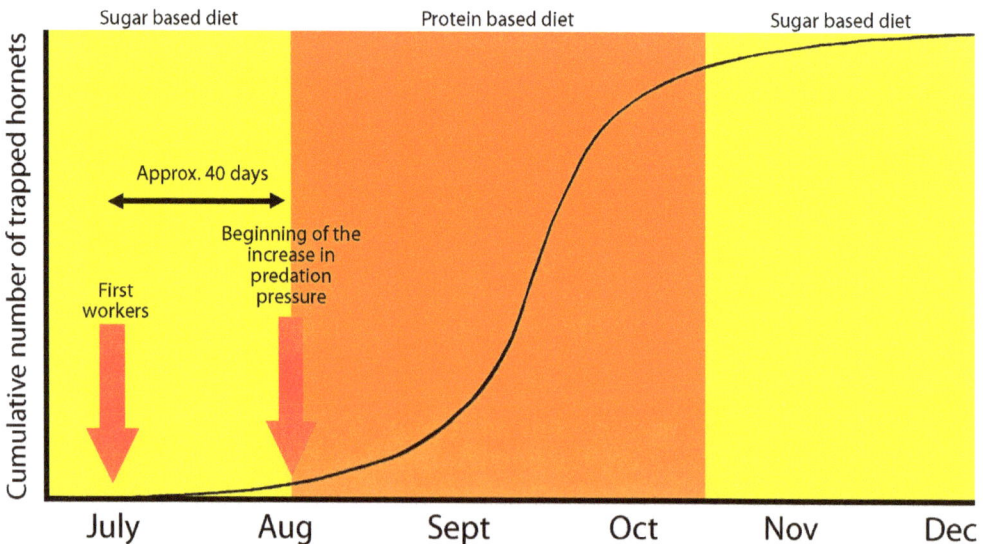

Graph after Monceau & Thiéry (2016): the three phases of trapping.

A single guard honey bee faces up to a yellow-legged hornet on a landing board of a hive in France. Photo by Sylvie Richart-Cervera / INRAE SAVE.

were more efficient in catching honey bees in the middle of the day (1–2 pm). Meanwhile, honey bee flying activity reduced over the course of the day. However, Diéguez-Antón et al. (2022b) found changing patterns of foraging by YLHs in Galicia, Spain. They found that in July and August, hornets avoided hunting in the middle of the day, but in September and October, they preferred hunting at that time. Temperature and humidity affected numbers of hornets at hives in Spain, with the most hornets found at temperatures between 19 and 25 °C with a relative humidity of more that 60%. Monceau et al. (2013c) found that temperature and humidity did not affect numbers caught at the apiary, but greater wind decreased the numbers trapped, suggesting that they are less likely to forage on windy days.

It is important to remember that YLHs are not specialist honey bee predators; they are generalists that take advantage of colonies of honey bees. Monceau et al. (2013c) found that the greatest number of honey bees were captured with nine hornets hawking in front of a hive: more hornets did not catch any more honey bees. The authors wonder whether the extra hornets may be acting as guards, perhaps defending this food patch from other hornets, or whether competition was occurring between hornets. Perhaps this effect is the result of foraging paralysis occurring (more on this in the next section).

In South Korea, where YLHs are most active in the morning, Choi (2021) recorded the duration and success of attacks on Western honey bees (*Apis mellifera*) around their hives, from July to October. He found that the hawking duration was 40–50 seconds before a hornet caught a honey bee from July to early August. Then, they became faster at catching honey bees: around 20 seconds at the end of August, and 10 seconds was recorded in September. After this, they gradually became slower once more, with 20 seconds at the beginning of October and 80 seconds at the end of October. He put the fast capture rates down to more experienced hornets in August, and the pressure of a rapidly growing colony to sustain. Subsequently, the slower capture rates in October were attributed to lower temperatures. The foraging success was 80.63% overall but exceeded 96% in September. This is a much higher success rate than that found for YLHs predating *Apis mellifera* in China (mean 30.6% success rate, Tan et al. 2007), so perhaps there is hope that some challenge against YLHs can be mustered by *Apis mellifera* over time. For comparison, Tan et al. found that the success rate for hornets catching Eastern honey bees, *Apis cerana*, was a mean of only 11.2%.

In France, a maximum of 20 hornets were observed by Monceau et al. (2014b) hawking in front of a hive. In Kashmir, Shah & Shah (1991) recorded an average of 24 *Apis mellifera* honey bees (Western

Yellow-legged hornets hunting at a hive in France (ringed in white). Two are walking on the hive body, perhaps looking for an isolated bee; three others hawk in front of the hive. Photo by Felix Gil.

Yellow-legged hornet dismembering a honey bee. Photo by Peter Kennedy.

honey bees) being caught in a 30-minute period, by *multiple hornets* in an apiary with three hives. By extrapolation, that would lead to 576 honey bees being caught over a 12-hour foraging day by the unknown number of visiting hornets, if the capture rate remained constant. It was not known in this study how many hornet nests were contributing to attacks on the apiary, but the number of individuals involved was enormous over time.

Scientists in France have looked at predation by individual hornets. At one apiary in France (six hives), with at least five hornet nests within a kilometre of the site, it was estimated by capture-mark-recapture methods that 360 individual hornets were visiting the apiary per day in early August, and made a couple of visits in a half-day. Over the 9-day experiment, there was a turnover in individuals, with only four individuals visiting on all 9 consecutive days. Two thirds of those initially marked returned at least once: overall at least half were captured more than once. Interestingly, the hornets did not prefer the largest or the smallest colonies: one hive in particular was preferred by the hornets, and the authors wonder whether this colony was the least aggressive. Diéguez-Antón et al. (2022b) found that the hive most likely to be attacked was the one not functioning very well, as measured by the honey bee colony's ability to maintain temperature in the hive. Finally, as seen in Jersey when timing hornet visits to bait stations, subsequent visits by the same individual became faster, suggesting that they became more efficient at finding their way (Monceau et al. 2014b).

Everyone seems to want to know how many bees an individual YLH kills in a day. In fact, there are few data available for working this out, and perhaps the calculation for consumption by a whole nest of hornets over a season is more meaningful, as done by Rome et al. (2021). However, here is a 'back-of-the-envelope' calculation. Using Monceau et al.'s (2014b) paper above, they note a couple of visits per half-day to the apiary by an individual hornet, so let's call that four visits per day. Then Choi (2021) found a foraging success rate of 86.67% (early August, South Korea): let's round that up to 87%. Using these numbers, the kill rate of an individual hornet in a day would be 87% of four visits = 3.48 bees taken by a hornet in a day. But then to find the actual effect on the honey bee population at the apiary, you need to scale this up by the number of hornets hunting in the apiary on a daily basis. Using Monceau et al.'s figure from above of 360 individual hornets visiting an apiary in a day (from at least 5 nearby nests), you arrive at 3.48 × 360 = 1,253 bees taken in a day by the surrounding hornet colonies.

I often come across the statement that 'An Asian hornet can kill 50 bees in a day', as if this was the average amount of honey bee consumption. But this seems unlikely and may be due to a misreading of an article by Perrard et al. (2009) in which a small captive colony was fed with honey bees, and the *whole colony* (15 individuals by the end of the experiment) consumed 25-50 honey bees in a day. It wasn't clear whether the hornets were restricted in the number of honey bees offered to them.

Yellow-legged hornets may enter an unguarded honey bee colony to steal pollen and honey, scavenge on recently dead bees and attack isolated ones, and take larvae (Shah & Shah 1991, Arca et al. 2014).

Honey bee defence

In the yellow-legged hornet's native range, the honey bees that they co-evolved with are a different species (*Apis cerana*) to those of Western Europe (*Apis mellifera*), and it turns out that the two species have very different defence capabilities.

Apis cerana have evolved several strategies to defend themselves against the yellow-legged hornet. First, they use an avoidance tactic to dodge the hornets. Normally, both *Apis mellifera* and *Apis cerana* approach the hive slowly and alight near the entrance, sometimes making 'sashaying' movements (slipping from side to side) as they come in. However, when *Apis cerana* is under attack from YLHs they speed up as they come in to land, making a straight dash through the most dangerous piece of air-space twice as fast as *Apis mellifera*, which come in slowly, making an easier target.

Poidatz et al. (2023) looked in detail at the flight performance of honey bees and YLHs in front of a hive in France, using 3D image-based tracking. They filmed 5,175 predator-prey flight interactions, out of which there were only 126 successful attacks by hornets: a 2.4% success rate. Higher hornet density led to a decrease in speed of bees leaving the hive, and an increase in speed of bees returning to the hive, together with more curved flight trajectories, suggesting hornet avoidance. Higher bee flight curvature resulted in lower hornet predation success. Interestingly, and corroborating results by Monceau et al. (2013c), hornet predation success improved with up to 8 hornets in front of the hive and then declined, perhaps due to competition among the hornets.

Apis cerana can recruit guard bees faster than *Apis mellifera* to protect the hive (Tan et al. 2007, but see Dong et al. 2023: *A. mellifera* seems to form a bigger bee carpet, but this might just provide more opportunities for *Vespa velutina* predation). They also have another couple of tricks to keep themselves safe. The first of these is 'shimmering' where between 10 and 200 workers gather around the hive entrance and shake their bodies from side to side. This looks unlike anything you have seen a Western bee do: a sort of fast jerk of the body, synchronised to some extent with the bees around, and sends an 'I see you' message to the hornets, keeping them about 15 to 30 cm away. The 'I see you' signal is thought to have evolved to benefit both prey and predator. The prey benefits if the predator backs off, and the predator benefits if it backs off from what could be a costly encounter when its prey is aware and ready to counter its attack. Tan et al. (2013) found that the 'I see you' signal was innate in *Apis cerana* exposed to yellow-legged hornets, emerging as a behaviour 48 hours post-emergence, whereas *Apis mellifera* did not produce a shimmering 'I see you' message. The bees shimmer more violently if a hornet approaches more closely (Shah & Shah 1991). Both honey bee species make 'bee carpets' — essentially a co-ordinated blanket of tightly packed bees around the entrance or on the landing board of the hive (a prerequisite for shimmering), and both species can produce a hissing sound (Arca et al. 2014).

In *Apis cerana*, if a hornet makes contact with a bee carpet, it can be pulled in by a couple of guard bees and is rapidly attacked by 30–60 bees, grabbing, biting, holding and stinging it. Initially a loose ball of bees is formed around the hornet, but gradually some bees retreat, leaving around 25 hard-core defenders that form a tight bee ball around the hornet for 30 minutes to over an hour, and the hornet dies from overheating, or possibly overheating combined with increased CO_2, humidity and honey bee stings. But such balling behaviour has a cost to the honey bee colony, and bees die during such an offensive (Tan et al. 2016).

In a study on heat-balling of the Asian **giant** hornet, *Vespa mandarinia*, by Eastern honey bees, *Apis cerana*, Yamaguchi et al. (2018) found that the high temperature produced by the bees (46 °C) dramatically shortened their life expectancy. They found that those bees that had been involved in a bee ball and exposed to this high temperature had a tendency to join a subsequent bee ball more aggressively, using the rest of their shorter lives in defence of the colony, rather than more and more bees in the colony becoming short-lived.

Gu et al. (2021) investigated the effects of stinging and temperature of the bee-ball in Western (*Apis mellifera*) and Eastern (*Apis cerana*) honey bees against the yellow-legged hornet in China. They found that dead YLHs found around and inside *A. cerana* and *A. mellifera* hives often had honey bee stings in their bodies, most commonly in the neck region. Through close observation, they found that 87% of hornets were able to remove stings in less that a minute; it may

Japanese honey bees (Apis cerana japonica) *forming a 'bee ball' in which two hornets* (Vespa simillima xanthoptera) *are engulfed and are being heated to death. Yokohama, Kanagawa Prefecture, Honshu Island, Japan. Photo by Takahashi (Own work, CC BY-SA 3.0).*

be that stings in the neck region are more difficult to remove. The investigators stung restrained hornets with guard bees and found that hornet survival decreased as the number of bee stings increased, even when the stings were quickly removed by the hornet. The combination of heat ball temperature (44 °C) and one sting (from *A. cerana* or *A. mellifera*) significantly reduced survival by 57%. Heating alone only reduced hornet survival by 25% after 3 hours. At 44 °C, the median time to hornet death was 14.5 minutes for hornets stung by *A. cerana* and 12.5 minutes for those stung by *A. mellifera*, perhaps due to the larger size, and therefore bigger stinger and greater amount of venom injected by *A. mellifera.*

Another mechanism of hornet death in a bee ball has been discovered in the southern European hornet *Vespa orientalis*: in this case the bees (a variant of *Apis mellifera*: *Apis mellifera cypria*) asphyxiate the hornet by restricting the movement of its abdomen, which cuts down its ability to breathe [hornets and wasps 'pant' by expanding and contracting their abdomens, pulling air in and out through a series of holes (spiracles) along their sides]. Both *Apis mellifera* and *Apis cerana* will also build propolis walls to narrow and block off entrances during attacks.

Dong et al. (2019) and Tan et al. (2016) studied a fascinating aspect of honey bee communication in relation to hornets in China. They investigated the 'stop signal', a vibrational signal given by a forager to a waggle-dancer to stop them recruiting more bees to a resource — either because of adverse food conditions, or danger from a predator. The signal is delivered by a bee butting its head against the body of a waggle-dancer and delivering a brief vibrational pulse generated by its wing muscles, which causes the waggle-dancer to freeze. They found that *Apis cerana*, the Eastern honey bee, could signal danger from two different hornets: the yellow-legged hornet, or the much larger and more dangerous Asian **giant** hornet (*Vespa mandarinia*). The stop signal for the **giant** hornet was 1.5 times louder than that for *Vespa velutina*, as well as being a higher frequency that was more effective in stopping the waggle-dance. Big hornets elicited more stop signals than small hornets. They also found that there were differences in signals generated by attacks on foragers compared with those due to attacks upon the nest entrance.

In another paper, Dong et al. (2018) looked at a different aspect of signalling between Eastern and Western honey bees and yellow-legged hornets: response by bees to hornet alarm pheromones. If a yellow-legged hornet is approached by a honey bee on a landing board of a hive, it will produce a small amount of venom that acts as an alarm pheromone. Eastern honey bees respond to this signal by creating a bee ball with around 25 bees. However, Western bees in the same situation responded more weakly, creating a ball with only five bees, and seemed to be responding more to the live hornet, than really recognising the alarm pheromone as an important signal to respond to.

Also in the Eastern honey bee, *Apis cerana*, Robinson (2013) noticed that absconding swarms of bees in a mango orchard in Thailand used various behaviours to rid themselves of pursuing hornets [in this case *Vespa velutina* and *V. tropica*], including flying short distances to effectively shed hornets. Interestingly, the swarms produced thin 'tails' (hanging down) and 'arms' (along the tree trunk or branch) that seemed to draw the hornets away from the main mass of bees. When the swarm made a short flight, some hornets landed on the place where the honey bees had been, examining the surface, and did not follow the swarm, allowing the bees to gradually shed the hornets. Whether the bees leave a specific scent to distract the hornets, or whether the hornets are only fascinated by the general bee swarm scent, is not known.

With this possible repertoire of behaviours, the Western bee does not do very well. They can form a bee carpet and hiss, but they rarely form a killing ball and do not shimmer. Western bees only weakly respond to the hornet alarm pheromone contained in the venom. All in all, their defence is pretty poor. The result is that, when given the choice between attacking *Apis cerana* or *Apis mellifera*, YLHs prefer attacking *Apis mellifera*, and will be three times more frequently found at *Apis mellifera* hives, taking *Apis mellifera* eight times more frequently than *Apis cerana* and being three times more successful at catching *Apis mellifera* than *Apis cerana* (Tan et al. 2007). In Kashmir, Shah & Shah (1991) observed on average one (range 0-2) *Apis cerana* caught in half an hour compared with an average 24 (range 12-44) *Apis mellifera*, by multiple hornets visiting the apiary.

This hive is no longer defended by guards or a bee carpet. The yellow-legged hornets are trying to get inside, but are stopped by an entrance only big enough for honey bees. Photo by Ana Diéguez-Antón.

Impacts on Human Activities

Effects on apiculture

Honey bee colonies are already weakened from several stressors. First by human-driven spread of pathogenic and pest organisms, including *Varroa destructor* (varroa mites that feed on honey bee body fat and spread virus diseases) and *Aethina tumida* (small hive beetle, not yet found in the UK); second by habitat loss, fragmentation and degradation; third by agricultural intensification (including pesticides); fourth by climate change; and fifth by invasive alien species (Neov et al. 2019). The introduction of *Vespa velutina* adds to the mounting barrage of stresses that honey bees have to deal with.

Leza et al. (2019) were able to measure the levels of activated genes and enzymes that indicate stress in bees from apiaries with or without attacking yellow-legged hornets. Those attacked by YLHs showed elevated levels of stress indicators, indicating the negative impact of the predator on bee health. In addition, this stressed state may make them more susceptible to other stressors, initiating a downward spiral.

The stress caused by being under attack by hornets comes during the critical pre-wintering season for honey bees when they need to get enough stores in to see them through the winter, and produce winter bees that will form the winter cluster around the queen. Colony collapse due to YLHs is multifactorial. First, there is the direct reduction of foragers by predation, which leads to less food coming into the hive. Linked with this, the queen lays fewer eggs and workers ultimately are not replaced fast enough. Meanwhile, a drop in the numbers of foragers also leads to young bees taking on foraging roles too early (they should be performing tasks inside the hive for the first 2–3 weeks of their adult lives). These precocious foragers spend less time foraging, complete fewer trips and take longer trips out. This results in even less food coming in, resulting in even more young foragers being recruited, creating a downward spiral for the colony, along with brood not being fed properly due to a lack of nursery bees, and even the possibility of cannibalism of brood (Perry et al. 2015). Finally, the presence of hornets in front of the hive can prevent foraging altogether. Monceau et al. (2017) found that 10 or more

This hive is well defended by a large bee carpet, but if foraging is badly reduced it is hard for the honey bees to maintain the colony. Photo by Ana Diéguez-Antón.

hornets hawking in front of a hive reduced the numbers of foraging bees to a handful. This behaviour is called 'foraging paralysis' where workers refuse to go out foraging, reinforced by stop signals coming from any foragers that do arrive back safely. During foraging paralysis, not only is no food coming in, (nectar and pollen), but water is not collected. 'Cleaning' flights are not made (bees normally defecate outside the hive), and sick or dead individuals are not removed, leading to more disease. Resins are not collected for making propolis, and fanning at the entrance is disrupted. All these things pile on more and more stress, but ultimately, Rojas-Nossa et al. (2022a) found that in addition to low food reserves, it was a deficiency in the nutritional levels of wintering adults and a reduced number of workers that led to winter deaths of the colonies they studied in Spain.

In a fascinating article by Dong et al. (2023), the authors studied colonies of Eastern and Western honey bees attacked by *Vespa velutina auraria*, in China. They found that Western honey bees severely reduced foraging and experienced higher hornet predation of foragers than the Eastern honey bee, *Apis cerana*, with Western honey bees experiencing more hornet attacks as the weeks went on. The hornet attacks resulted in reduced egg production by the queen (within 1 week, a decline of 16%; a decline of 96% over 6 weeks), fewer pupae and fewer workers (decline of 8% in first week), although the major decline in workers only kicked in after 3 weeks of attacks (*A. cerana* experienced a small decrease in workers in week 5). Western honey bee colonies suffered an average decrease of 88% of workers over 4 weeks. By the end of the experiment (6 weeks), all the Western honey bee colonies (30 of them) were dead. None of the *Apis cerana* colonies died. The Western honey bee colonies had, on average, 4 times as many guards at the nest entrance (in contrast to the work of Tan et al. 2007, who found higher guard recruitment in *Apis cerana*), but the hornets could *collectively* capture around 100 Western honey bee workers at the entrance to a hive in a day. The authors suggest that, in addition to direct predation, predator-induced stress may contribute to Western honey bee colony decline.

In Europe, although honey bee colony death has been seen in full-sized colonies during hornet attacks, many more colonies are lost over winter due to small, stressed colonies with not enough stores. When foraging paralysis occurs, it is even easier for hornets to pick off the few individuals who do fly (Requier et al. 2018). The authors suggest merging small colonies and supplemental feeding with bee bread (pollen and honey) with low hornet pressure, and apiary trapping and muzzles with high hornet pressure, along with supplementary feeding both pre-winter and winter: something that French beekeepers have become used to doing.

As we have seen, honey bees are a significant percentage of YLH predation, resulting in some summer colony deaths, and more winter losses of colonies in those areas where the hornets have become well established. It is not yet known whether *Vespa velutina* is also capable of transferring diseases to honey bees: it seems that it can host Israeli acute paralysis virus (IAPV), which affects European honey bees in France and China (Blanchard et al. 2008, Yanez et al. 2012). Dalmon et al. (2019) found viral RNA in French YLHs: deformed wing virus

(DWV-B was predominant), Kashmir bee virus, acute bee paralysis virus, black queen cell virus, aphid lethal paralysis virus, bee macula-like virus and mokuvirus were all present. Some of these viruses were found in the brain and muscles, suggesting possible replication in the hornet. One hornet was found with deformed wings that could be due to deformed wing virus, which was found in all samples. The authors did not know whether these viruses pose a threat to honey bees, or whether they could be used as a control measure against YLHs.

When investigating by-catch in kill-traps set out for hornets in northern Spain, Sánchez & Arias (2021) found that 5 of the 286 hornet specimens recovered (out of 12,835 other killed insects) had *Varroa* mites attached to them, clinging to the under/side parts of their abdomens.

The data from France on honey bee colony losses appear patchy, with some areas reporting greater losses than others. There are very few data available on actual colony losses, but the figure of 30% destroyed or weakened is generally referred to (Monceau et al. 2014a). In a study mentioned earlier (Monceau et al. 2014b), the study apiary consisted of six hives, and out of these one colony was completely destroyed while the colony sizes of the other five hives were halved. Shah & Shah (1991) found that *Vespa velutina* limited colony development by persistent predation of adult bees in Kashmir. Colony losses are recorded in France (beekeepers have to register their honey bee colonies, and surveys of colony losses are well reported); the problem is that it is difficult to put losses down to any one cause when pests, diseases, lack of forage, insecticide use and especially weather can all cause colony losses. In Portugal, Nave et al. (2022) put losses down to weather first, *Varroa* second, and YLHs third. In northern Spain, Ferreira-Golpe et al. (2020) found that production was affected more by weather than the presence of hornets, but the level of management, particularly targeted feeding, during sustained attacks from YLHs also had an effect on honey production levels.

Diéguez-Antón et al. (2022b) also found, in northern Spain, that the level of pressure hornets exerted on apiaries was influenced by weather, the number and sizes of local hornet nests, honey bee colony strength, management practices, and control measures. Although 42 nearby nests were destroyed over the 2-year observation, hornet pressure remained high, and extended for more than 6 months each year. As part of their investigation, they measured the levels of thermoregulation in the hives and discovered that the key factor in colony survival was the ability of the colony to regulate temperature within the hive. Therefore, management practices that strengthened colony health, including feeding, were essential for keeping colonies going in the face of constant attack.

Anecdotally, some French beekeepers have reported 40% losses and worse but, again, losses are patchy (Kevin Baughen, pers. comm.). Decante (2015) reported losses of up to 60% in colonies at the end of winter in Alpes Maritimes, France. Laurino et al. (2020) mention the loss of almost 50% of colonies in some European regions, and losses of between 30% and 80% in south-west France.

Loss of colonies puts stress onto beekeepers too, (Monceau et al. 2014a), with loss of income for professional beekeepers, and some amateur beekeepers simply giving up, as was the case when varroa arrived in the UK. As well as honey bees and the pollination services they provide, the products that come from farming bees (honey, wax, pollen and propolis) are reduced when colonies are lost (Ferreira-Golpe et al. 2018).

Fruit production

As we have seen, yellow-legged hornets affect honey bees and therefore beekeepers, and possibly horticulturists who rely on honey bees for crop pollination. Fruit growers in France and Spain have been affected by hornets feeding on ripe fruit and pickers getting stung: there is a cost to 'red fruit' farmers, and to others who work with trees. In Spain, some crops are being grown in netted tunnels to try to avoid damage to fruit (Ana Diéguez-Antón, pers. comm.)

Nave et al. (2022) reported some attacks on grapes, plums, figs, pears and apples in Portugal, in 2019 and 2020, with up to 50% of fruits affected, but it was not possible to differentiate between yellow-legged hornet and European hornet attacks.

Human health — stings

Another impact of yellow-legged hornets on humans, apart from economic and ecological, is the harm that can occur if someone gets too close to a nest and gets badly stung. As we have seen, yellow-legged hornets thrive in urban settings and can reach high densities; when this is the case, more people are going to accidentally disturb nests while going about their normal activities (cutting hedges, fixing roofs, reading utility meters, etc.). People will have to learn to be more careful. Apart from actual harm, there is also fear of these insects, which can be fuelled by the media, leading to pointless worry and the killing of many other insects through misidentification.

Wasps, hornets and honey bees can all deliver painful stings, and unfortunately for some people who are allergic, a single sting can lead to a severe reaction and even death. The yellow-legged hornet,

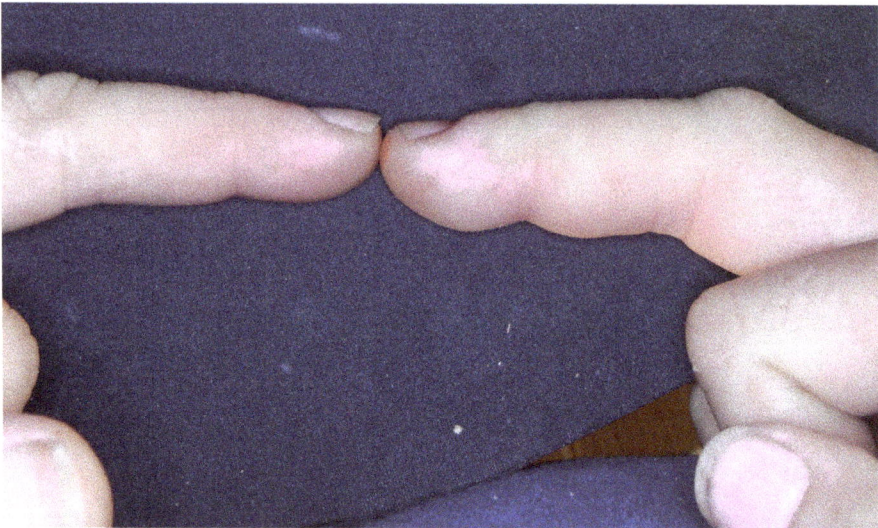

Finger swollen due to yellow-legged hornet sting. Stings are very painful. Photo by Chris Isaacs.

being quite a large insect, can deliver a deep and painful sting and can inject more venom than a smaller wasp. Those who have been stung usually say that it is much more painful than a bee or a wasp sting (Alastair Christie, pers. comm.).

Honey bees have barbed stings, which get trapped in the elastic skin of mammals so effectively that when they pull away, the stinging organ is ripped from their bodies and the bee goes on to die. Wasps and hornets have smoother stings, which allow them to sting multiple times without getting them caught in the skin of their victim.

The sting of a worker YLH is around 3–3.5 mm long (personal observation; Gu et al. 2021) and the specialist suits that are used by professionals who destroy and remove nests are 5 mm thick or more. This is achieved by padding or special materials that are breathable but still maintain that essential thickness.

According to a French pharmacologist who specialises in toxins at the anti-poisoning centre in Angers, Gaël Le Roux, the venom of the YLH is a little less toxic than that of its European counterpart (*Vespa crabro*), but can still trigger a very serious allergic shock (Troadec, 2015). Scientists in China have analysed the components in *Vespa velutina* venom and found toxins that affect the integrity of blood and how it clots, and neurotoxins (which affect the nervous system) that are responsible for the toxic reactions and allergic effects (Liu et al. 2015). Being badly stung can result in acute kidney injury and renal failure, and 20 stings or more leads to increasing mortality rates (Carriazo & Ortiz 2022). Liu et al. (2022) suggested a wasp sting severity score that could guide the application of life-saving plasma exchange.

It is possible to have venom immunotherapy against various hymenopteran stings, a treatment that injects small amounts of venom over a long period (up to 5 years) to prevent future serious allergic reactions. Supplies of *Vespa velutina* venom will need to be regularly obtained for venom immunotherapy treatment when yellow-legged hornets establish in this country (Feás et al. 2022).

Yellow-legged hornet stings. Stings are 3–3.5 mm long. Photo by John de Carteret.

Becoming more of a problem is the squirting into eyes of a liquid projected from the back ends of yellow-legged hornets. It is not known exactly what this liquid is (venom or something else?), but it irritates 'like getting washing-up liquid in your eye', resulting in conjunctivitis. It mainly affects those who remove nests. Bob Hogge, in Jersey, described it during the taking down of a large live nest, where he could see hornets landing on the veil of his hornet suit, curving their abdomens around as if stinging, and releasing a stream of liquid towards his eyes. He was wearing ordinary glasses, but this isn't enough eye protection. Nest removers have tried safety specs, but these steam up too quickly, so people are turning to hard, clear plastic visors worn between the face and the veil to give enough protection without steaming up. I give more information about safety equipment needed when dealing with hornets in the chapter 'Nest destruction'.

Alastair Christie's face visor against fluid squirted from yellow-legged hornets. He has added extra ventilation holes along the sides and under the chin.

Laborde-Castérot et al. (2020) looked at 29 cases of this liquid squirting into eyes in France. Most cases (80%) were in professionals dealing with hornet nests. Generally, the symptoms resolved quickly (hours or days) after decontamination with water or saline solution, but one suffered from swelling around the eye for 6 days and another had head pain for 20 days. They noted that although the symptoms could be consistent with venom, the venom reservoir is only 1 mm in length, and so not able to hold much liquid: perhaps the liquid originates from the intestinal tract, with or without venom.

The implications for public health and the worries about honey bee losses are at the forefront of most grass roots Asian hornet organisations in France. Already several people have died in France and Spain due to YLH stings, but according to a study by de Haro et al. (2010), there has been no increase in the number of hymenopteran stings in areas of France where *Vespa velutina* has become established. Meanwhile, Pretre et al. (2022) state that *Vespa velutina* has become the most prevalent cause of hymenopteran anaphylaxis in north and north-western Spain.

There seem to be two particularly dangerous situations when it comes to YLH stings: the first occurs when the victim is predisposed in some way to a serious outcome, such as having an allergy to stings, or a history of heart problems, in which case even a single sting can be serious, as with other species of hornet or wasp. The second situation is when the victim receives multiple stings, which only occurs when they get too close to a nest. If you accidentally get within around 5 m of a nest (de Haro et al. 2010, 3 m: Choi 2021b), a full-scale attack may be triggered with many or most of the hornets in the nest being recruited to see you off. Choi (2021b), working in South Korea, found (using hornet-suited volunteers wearing black wigs) that worker hornets showed defensive behaviour when a person approached within 3 m of the nest, and that the attack rate was higher if the person conducted a large arm swing. When someone touched the nest and then walked away, the number of hornets attacking did not significantly decrease until they were 5 m away, and more than halved at 10 m. A tenacious minority (one or two workers) kept attacking up to 300 m. However, if the person ran, the number of workers attacking reduced to around one after 10 m. Therefore, he recommends running away from an attack while trying to protect your face. A wide-brimmed hat prevented about 75% of stings to the head. Kim & Jung (2019, also South Korea) found that hornets preferentially attacked black, their defence boundary was approximately 15 m, but they recognised waving disturbance at 10 m from the nest. Finally, Choi et al. (2021) presented coloured and hairy/hairless bundles to nests, again finding that black and then brown triggered the biggest defensive reaction; the attack was longer for the hairy bundle; and there was no defensive response to human conversation or loud music.

Yellow-legged hornet venom acts as an alarm pheromone, which attracts more workers and causes them to attack (Cheng et al. 2017; Thiéry et al. 2018). You don't even need to be stung to trigger this response: if the hornet is alarmed enough it can exude some venom from its sting, and the volatile components released will stimulate others to attack.

Martin (1995, 2017) says that in many parts of Asia *Vespa velutina* has a particularly aggressive nature around mature nests, and is greatly feared. In Indonesia, Vietnam and China the species is thought of as aggressive (de Haro et al. 2010). Indeed, Liu et al. (2015) describe it as 'powerful and deadly...the most aggressive and fearful species in China.' and they go on to say that it has become a 'severe public health concern', with 2013 being a particularly bad year. This raises several questions: is the species more aggressive in its native habitat? Will the invasive YLHs in Europe become more aggressive over time (perhaps due to nest density)? Or is the aggression seen in China during recent years simply the result of closer human-hornet contact through massive development projects, which have turned some areas of countryside into towns and cities?

As mentioned before, yellow-legged hornets do extremely well in urban environments (Monceau & Thiéry 2017, Choi et al. 2012), and humans are more likely to accidentally get too close to a nest in this situation. Of secondary nests in south-western France, 78% were on natural structures (mainly trees), Franklin et al. (2017). The Association Action Anti Frelon Asiatique (the AAAFA), a French grass-roots organisation, lists the dangerous places at ground level or only a metre high that nests have been found in, including under manhole covers, inside gas and electricity meter boxes, in hedges, arbours and fruit trees, and various building cavities. They suggest that an attack can be triggered by strong vibrations and noise (such as the use of some garden machinery), and recommend always checking suspect places (even if they aren't actually showing hornet activity) by tapping with a long-handled broom, while being ready to run!

The costs of combatting yellow-legged hornets

Dealing with the effects of the yellow-legged hornet costs money, whether it is for control or health treatment, and there are economic losses due to reduced production in apiculture and horticulture, and the possible impact of diminished ecological services.

Barbet-Massin et al. (2020) looked at the economic cost of controlling the yellow-legged hornet, using data from France and extrapolating this via modelling to other countries. They estimated that the cost of nest destruction in France between 2006 and 2015 was €23 million. In yearly terms, they estimated that costs could reach €11.9 million in France, €9.0 million in Italy and €8.6 million in the UK if the species fills the areas that are climatically suitable. Despite such high costs, they recommend increasing nest destruction in order to avoid the further costs to apiculture and pollination services.

In a more recent paper by Requier at al. (2023), the authors estimated the hornet-related risk of bee colony mortality, the economic cost of colony loss for beekeepers and the economic impact of bee colony replacement compared with honey revenues, at an estimated density of 1.08 hornet nests per square kilometre. They estimated that 2.6–29.2% of bee colonies were at risk of being lost each year due to yellow-legged hornets. These losses correspond to an economic cost of between €2.8 million and €30.8 million in France annually. For beekeepers, this cost may be equivalent to up to 26.6% of their honey revenues, due to having to replace bees. Even in a very low predation scenario, the cost would be €2.8 million per year, with beekeeper costs of 2.4% of honey income. The costs incurred by beekeepers could be three times higher than the control costs. Ongoing high bee colony losses mean that, worldwide, beekeepers need to increase their bee colonies by about 130% each year, and charge more for pollination services to farmers. They also point out that their calculations don't take into account the impact on wild pollinators and ecological costs, which could be orders of magnitude higher, and not measurable by the same standards.

Garcia-Arias et al. (2023) surveyed beekeepers in Portugal, northern Spain and France (378 beekeepers) to estimate costs due to yellow-legged hornets. They found that the total costs represented around 10% of the estimated production value.

Yellow-legged hornet worker gathering bait. This bait station has plastic netting which avoids the hornets getting too sticky. Photo by Angus Deuchar.

Part II
Context

Spread of the Yellow-Legged Hornet

The yellow-legged hornet arrives in Europe

The first confirmed recording of a yellow-legged hornet in France was made in Nérac, Lot-et-Garonne (inland in south-western France) in 2005, by Jean-Pierre Bouguet, an amateur entomologist (Haxaire et al. 2006). When another three hornets were found the following spring, it was suspected that these were probably foundresses too, and it is possible that one or more colonies could have been established in the area for at least a year, putting the invasion at 2004, possibly even earlier. Meanwhile, a nearby bonsai producer had noticed brown hornets flying in the summer of 2004, which he recognised from a trip to China not long before. It is quite likely that this is the origin of the French invasion: a fertilised female could have easily survived the month-long journey by boat from Yunnan, China, hibernating during winter in a box of pottery (Villement et al. 2006). It seems now that a single, multi-mated queen was responsible for this spread through Europe, which still continues (Arca et al. 2015).

Vespa velutina naturally occurs in Asia from northern India to eastern China, Indo-China and Indonesia (Carpenter & Kojima 1997), and genetic marker comparisons do support the idea that the French population came from eastern China (Jiangsu/Zhejiang areas) (Arca et al. 2015).

Going to press (2024), all Departments in France have been invaded, along with northern Spain (2010; years in brackets are first discoveries), two thirds of Portugal (2011), Belgium (2013), Mallorca and Germany (2015: but now eradicated in Mallorca), Italy and the Channel Islands (2016), the Netherlands (2017), Luxembourg and Switzerland (2020), Ireland (2021), Hungary and the Czech Republic (2023) (Requier et al. 2023), and Slovakia and Austria (2024). In the UK, nests and individuals have been found and destroyed since 2016.

Separate dispersals from its native region have spread the yellow-legged hornet to South Korea, and thence to Japan (Island of Tsushima, where it is now established, and nests found on Kyushu,

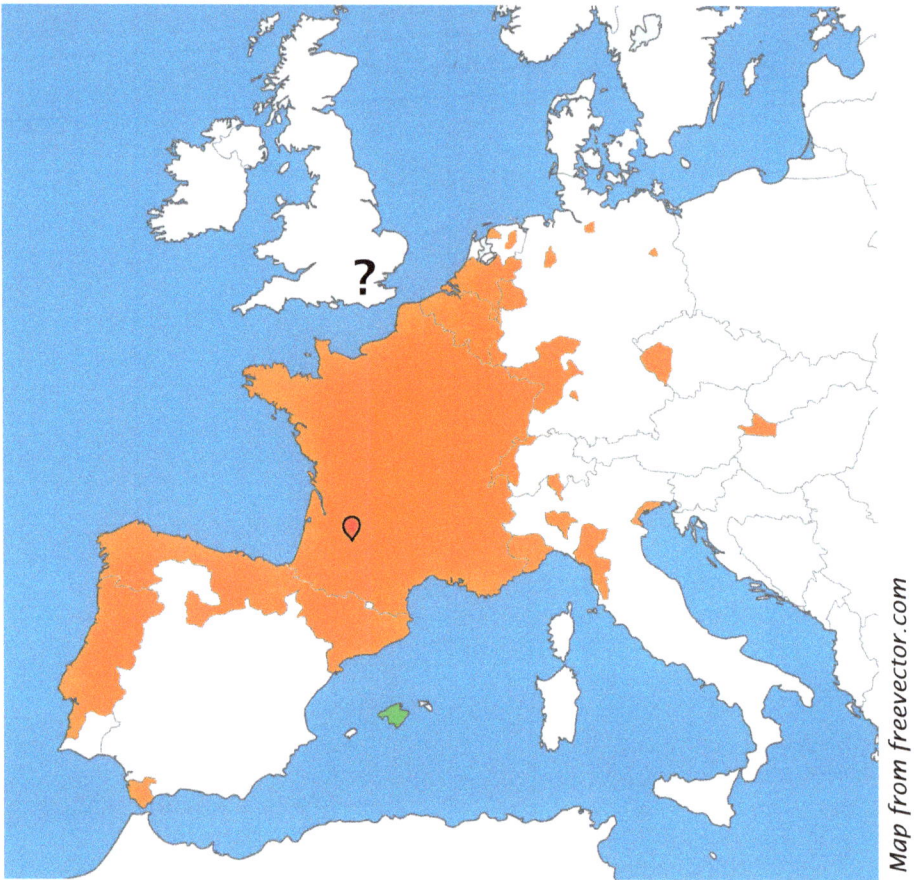

Map from freevector.com

*Map of Europe, showing colonisation, in **orange**, by yellow-legged hornets. The initial introduction into France is marked with a **red pin**. Mallorca, where eradication has been successful, is **green**. We still don't know whether yellow-legged hornets have established — hence the question mark.*
This map is based on a map of yellow-legged hornet spread in Europe maintained by Quentin Rome at the Museum National d'Histoire Naturelle (MNHN: National Museum of Natural History in France). It is up-to-date at time of publishing. For a current map, please go to their website:

http://frelonasiatique.mnhn.fr/home

Honshu and Iki Islands) (Ikegami et al. 2020). And another long-distance journey, probably by cargo ship, resulted in 5 nests being found in Savannah, Georgia, USA in 2023. These nests were first generation (they shared the same maternal lineage), so the original queen probably arrived in 2022, and could have come from Japan, South Korea or China (Lewis Bartlett on Bob Binnie interview, 2023). Fourty-seven nests were found in 2024.

Country	First detected	Status
France	2004	Established
Spain	2010	Established
Portugal	2011	Established
Belgium	2011*	Established
Mallorca	2015, 2021	Both incursions eradicated
Germany	2014	Established, patchy
Italy	2012	Established, patchy
Channel Islands	2016	Established
UK	2016	Incursions, establishment unknown
Netherlands	2017	Established
Luxembourg	2020	Established
Switzerland	2020	Established
Ireland	2021	Single individual
Hungary	2023	Incursion
Czech Republic	2023	Incursion
Slovakia	2024	Incursion
Austria	2024	Incursion
South Korea	2003	Established
Japan	2012	Established, Tsushima Is.
USA	2023	Incursion, establishment unknown

Table showing countries invaded by the yellow-legged hornet, when first detected and the status of each invasion. Belgium: a single male found in 2011; first nest found in 2016.*

Speed of spread

The speed of spread through Europe has been fast. Indeed, the expansion through France has proceeded at around 78 km per year (the pioneer population possibly achieved 30 km in the first year; Robinet et al. 2016) with some long-distance jumps from the colonisation front of around 250 km, which are most likely due to human transportation (Rome et al. 2009). This is four to eight times faster than in South Korea, where a similar invasion of *Vespa velutina* began around the same time (2003) (estimated spread 10–20 km per year, Choi et al. 2012; 12.4 km per year, Jung 2012).

In Spain, the French population came across the Pyrenees in 2010, and then also suddenly turned up on the Atlantic coast, in the far north of Portugal in 2011, near a paper mill, probably arriving in a timber shipment from France (Verdasca et al. 2022). The following year, 2012, they were found in two separate places in Galicia simultaneously: right down in the south on the Atlantic coast, not too far from the Portuguese border, and in the north, on the coast. Rodriguez-Flores et al. (2019) suggest that these two probably different populations aided the very rapid spread of the hornet through Galicia. By 2014, these populations had spread along the entire north coast of the Iberian peninsula.

Meanwhile, the yellow-legged hornet moved rapidly down through Portugal. Verdasca et al. (2021) found that the southward spread along the Atlantic coast was at 45 km per year, while spread in an easterly direction was slower, at 20 km per year.
In the rest of continental Europe, the spread out from the French point of origin has been steady and inexorable, with a slow movement into Italy. Bertolino et al. (2016) found that the invasion front from France was advancing into Italy along the coast at around 18.3 km per year; this has since been reduced to around 3.5 km per year as a result of several control measures. There have been some long-distance jumps within continental Europe: Portugal, as mentioned, Hamburg in the north of Germany, and recently Hungary as well as the Czech Republic. And there have also been long-distance jumps to Mallorca, Ireland and the UK, facilitated by human transport.

What does natural spread look like if there is no hitch-hiking on human transport? This is an important question if we want to know where to implement control measures such as surveillance, or how to model spread throughout a territory. There have been a few studies that have looked at the distribution of nests at early stages of colonisation. For example, when Leza et al. (2020) looked at the spread of nests in Mallorca, in 2016, they found that the mean rate of spread from the first nest was around 5 km (maximum 9.57 km), while in 2017 it was around 4 km (maximum 11.40 km) from the second wave of nests. In Bertolino et al.'s (2016) study of nest distribution in Liguria, northern Italy, they found core areas or clusters that accounted for 90% of nests in 2015; the rest were widely spread single nests. Lioy et al. (2019), looking at the same data for Liguria, found that 95% of nests were built within 1.4–6.2 km of previous nests. It is a common feature in the dispersal of organisms generally, that a large proportion end up within a certain range, while a small proportion go much farther.

So, what has been facilitating or hindering the spread of the yellow-legged hornet in new territories? In general terms, as far as climate goes, the whole of western Europe seems conducive to invasion by YLHs, including Ireland and as far north as Denmark. One model (Villemant et al. 2011a) predicts climatic suitability right up into Scotland and coastal Norway, based on the range of its native territory, but when combined with data from invaded areas, the predicted range is a little more limited. Climate is the most important factor to consider, and especially the availability of water and mild enough temperatures. Balmori (2015) superimposed *Vespa velutina* distribution in 2015 with various climatological maps, and found that an average spring temperature between 12.5 and 15 °C, approximate average winter temperature of 10 °C; maximum 20–30 days of frost per year and maximum 10 days frost in spring; average annual precipitation greater than 700 mm and average rainfall in March and April above 100 mm were all factors that matched the hornet distribution. If you squint at his maps, average annual precipitation seems to match the present (2023 season) hornet distribution best. The hot, dry interior of Spain does not look very favourable for yellow-legged hornets, but would seem to be more suitable for *Vespa orientalis*, the Oriental hornet.

I mentioned that the spread of the hornet south through Portugal was faster than its spread inland towards the east. Verdasca et al. (2021) put this down to a more abrupt transition to the drier more Mediterranean climate heading into the Iberian interior, plus the hitch hiking advantages of the more established transport routes running north-south in the country.

Climate change is, of course, another factor to consider in predictions of spread. Barbet-Massin et al. (2013) looked at climatic suitability for yellow-legged hornets in 100 years' time, based on changing patterns of temperatures, precipitation and seasonality, and found that their predicted range could extend east into central and eastern Europe — to countries which at present have the highest densities of honey bee colonies. In a more recent study, Verdasca et al. (2022) found that there is nothing to stop them invading those areas now. Indeed, this seems to be what is happening.

As Martin (2017) points out, however, although *Vespa velutina* has a massive natural range, this range is occupied by a complex of 12 colour forms: these different colour forms are probably acclimatised to different conditions within this huge area.

This brings us to ecological niche shifts. The niche of a hornet population is the specific habitat (including other organisms) that the population occupies. So if a fertilised queen hornet takes a long-distance trip on a log-truck from central France, and ends up in Portugal, is it now occupying a different niche? Verdasca et al. (2022) looked at this question and came to the conclusion that yes, there has been a niche shift from the French population, and the Portuguese population is also thriving under conditions different from the native range of the hornet in Asia. Why is this important? Because when trying to model how the hornet will spread long term, you need to know whether it can only thrive in the niche it first established in, or whether it can adapt to new areas. Unfortunately for us, the requirements of yellow-legged hornet are widely available in Europe, meaning that is has a massive potential range.

Within the broad climatic areas suitable for establishment of the yellow-legged hornet, there are certain landscape features that seem attractive, and others that might serve as barriers to their progression. Altitude appears to restrict distribution in both its native range and in Europe. In a 2016 study, Bertolino et al. showed that all nests found in Ligura (north-west Italy, coastal strip) up to that point had been at an altitude below 620 m above sea level. Further north, a nest was found in Piedmont at 906 m above sea level, but the next year the hornet was not present in the area. Indeed, the distribution of the invading front seemed squeezed into a narrow coastal strip between the sea and the Alps. Looking at the same area, Lioy et al. (2019) found that 95% of nests were below 521 m above sea level. In Spain, Rodriguez-Flores et al. (2019) found that the average yellow-legged hornet count in traps decreased above 200 m, although they still caught a few above 600 m. Indeed, Diéguez-Antón (pers. comm.) mentions that nests at 1000 m and above are becoming more common in Galicia, northern Spain (data from Xunta de Galicia). Carisio et al. (2022) considered 750 m above sea level to be the upper limit for *Vespa velutina* to nest in Mediterranean areas. In South Korea, Do et al. (2021) trapped individuals on mountains at an altitude of 1100 m above sea level but thought that their nests were probably lower down; however, they did not think that snow or cool temperatures slowed their spread.

In an interesting study that looked at a whole raft of environmental factors affecting the distribution of *Vespa velutina* in Mallorca (Herrera et al. 2023), steeper slopes and stability of the local climate in terms of temperature and precipitation were the most important factors for nest sites. The element of steeper slopes is fascinating. Steep wooded slopes might allow morning warming for the nest in the top of a canopy, if the slope is pointing the right way, and offer a spread of visual cues without the hornet having to gain height above the surrounding canopy first.

Availability of water and progression along water courses has also been described in many studies (e.g. Monceau et al. 2012, Rodriguez-Flores et al. 2019, Godhino & Nave 2020, Carvallho et al. 2020, Caragata & Montesinos 2020), but this may only be a limiting factor in drier places (Rojas-Nossa et al. 2018). Several factors may naturally correlate here in intensively agricultural areas: river and stream valleys in rolling countryside are generally too steep to be cultivated and are therefore wooded, and the same can be said for the incised valleys in Jersey. Even in flat places, wild areas with more potential nesting sights (and possibly more prey) are often left along river banks.

Verdasca et al. (2021) looked at the distribution and spread of *Vespa velutina* in Portugal. As well as determining the speed of spread, they discovered that there were more nests close to motorways than would be expected by chance, and they suggested that early monitoring should include zones around motorways as well as ports. Motorway transport that may particularly spread YLHs could include timber, especially unprocessed wood, and landscape/garden trades.

In South Korea, the spread of the yellow-legged hornet may have been slowed by competition with the six species of native hornets already there (Villemant et al. 2011a). There is much competition and predation between hornets, and a fairly clear pecking order. Kwon and Choi (2020) used gladiatorial combat to determine aggression in relation to body size in the six species. The body size hierarchy goes from *simillima* and *velutina* as the smallest, but very similar, to *dybowskii*, to *crabro* and *analis* (again of similar size) to the undoubted giant, *mandarinia*. In terms of aggression between the hornets, the order from least aggressive to most was *simillima*, *velutina*, *crabro*, *analis*, *dybowskii* and finally *mandarinia*. From this analysis, you can see the hopes of Europeans that *Vespa crabro* might interfere with the spread of *Vespa velutina*, but it would probably need a larger population of *V. crabro* than we have: it certainly hasn't stopped the spread through France. In Italy, Carisio et al. (2022) could not discern competition between the two species, but this may have been because *V. velutina* was not abundant enough for competition to occur. They also noted that competition between the two may be avoided by *V. crabro* emerging later in the spring, and preferring higher regions. Because the rate of spread of the YLH into Italy was so much slower than through France, the authors did wonder whether the healthy population of *V. crabro* in northern Italy might have acted as a biological barrier.

Meanwhile, on Tsushima Island (Japan), about halfway between South Korea and Japan, Ikegami et al. (2020) tried to determine whether competition was occurring between the six native wasps (including four native hornets) and the newcomer, *Vespa velutina*. To do this, they had to remove the effects of the environment on their distributions, and found that the four native hornets competed with and/or predated one another. That *Vespa velutina* was able to expand into this territory shows that it can hold its own to some extent, but its slow expansion to the south of the island may be due to a biological barrier from other hornets, especially *V. simillima xanthoptera* and *V. analis*. Although the northern giant hornet (aka Asian **giant** hornet, *Vespa mandarinia japonica*) was found to coexist with *V. velutina*, this giant hornet could also be a barrier to the expansion of the yellow-legged hornet.

Of course the sea isn't a barrier when hornets hitch rides on ferries (there were four sightings on ferries in 2023: Nigel Semmence's talk at the National Honey Show, 2023) and among cargo in holds, or on lorries, but how much water can a yellow-legged hornet cross without running out of steam? Sauvard et al. (2018), set up flight mill experiments to explore flight capabilities in terms of foraging, so hornets were rested and fed every hour, making it impossible to judge continuous flights longer than that. Mated queens emerging in the spring have not been tested on a flight mill, but they do migrate distances of at least 10–20 km (I'm basing this on average yearly spread distances). However, this would involve plenty of stops to rest and refuel. Flying over water non-stop is a different matter, but could be made much more possible with a strong wind at your tail and/ or rests on ships (as we know, the Channel is a very busy shipping route).

That hornets are able to fly from mainland France to Jersey seems clear (generally, more nests are found on the side of the island closest to France): Jersey is only about 12 miles (19 km) from France, and there are intermediary reefs that could be staging posts. If crossing the channel to get to Kent, the minimum distance over the sea is around 22 miles (35 km). It could be that this is at the limit of their continuous flight cababilities, unless helped by a strong wind from the continent. In October 2018, two male yellow-legged hornets were found in Dungeness, Kent and were assumed to be 'blow-ins'.

To summarise, the spread of the yellow-legged hornet through Europe has been fast, especially in France. Hornets can hitch rides and therefore can turn up virtually anywhere. They seem to be limited in the altitude that they will build nests (mainly below 750 m, but in Galicia where the tree-line is higher, they are nesting at higher altitudes), and are discouraged by very dry Mediterranean climates. They like a stable temperature with enough rain and favour steep wooded slopes along river valleys for nest building, but can and do build nests anywhere the climate allows, including in suburban and urban areas. The European hornet does not seem to act as a biological barrier to the progression of the yellow-legged hornet.

The yellow-legged hornet arrives in the UK

For an up-to-date map of sightings of yellow-legged hornets in the UK, with brief details, go to the British Beekeepers' Association (BBKA) website, scroll down the main page to 'Asian Hornet', click, then scroll down to 'Asian Hornet Sightings and Incursion Map. There is also a rolling update for YLH sightings on the BeeBase website (www.nationalbeeunit.com), and you can sign up to the BBKA Asian Hornet Team Announcements WhatsApp group (this is purely announcements: no chat).

The first reported sighting of a yellow-legged hornet in the UK was at an apiary near Tetbury, Gloucestershire, by a beekeeper, in mid September 2016. The nest was located and destroyed a couple of weeks later by the National Bee Unit (NBU). It was found near the top of a 55-foot cypress tree, and the vast majority of the workers were observed foraging within 700 m of the nest (92 of 94 observations within 700 m; another two observations 1.15 km away, feeding on ivy), which fits with other observations on foraging ranges (e.g. Poidatz et al. 2018a). Once analysed, the nest was found to contain 70 adults: 57 females (workers) and 13 males. Interestingly, the males were diploid, therefore likely to be sterile. There were also eggs, larvae and pupae. Genetic analysis showed that the queen had only mated once, which suggested low nest density, and the genetic markers were consistent with these insects being closely related to the French population: definitely part of the same European invasion (Budge et al. 2017). To check that this nest was the only source of YLHs, more than 220 field inspections of apiaries and flowering forage sites were conducted within a radius of 17.5 km from the nest. No further hornets were found.

Since that first discovery of a nest, there have been a few nests found each year, and also individual hornets that didn't seem to be connected with a nest and which were assumed to be accidentally transported from the continent (see Table). Until, that is, 2023, when sightings and nests destroyed increased dramatically. In 2023, 72 nests were found in 56 separate locations. The vast majority were in Kent. Seven nests were found in a particularly gruelling environment, Capel-le-Ferne, an SSSI of 782 acres of incredibly impenetrable, steep, rough undercliff full of brambles, thorns, buddleias and trees. So steep that even the hornets had to zig-zag to gain altitude. One nest at a height of 38 m needed a tracked cherry-picker at full height (24 m) plus an 18-m pole to reach the nest.

Year	No. nests found	No. single hornets found	Notes
2016	1	2	Size: 23 cm
2017	1	-	Size: 27 cm
2018	4	5	Size: 15, 19, 18, 18.5 cm
2019	3	2	
2020	1	-	Covid
2021	2	-	Covid
2022	1	1	
2023	72	Unknown	Nests found in 56 locations. Some full-sized nests found

Table showing UK incursions from 2016–2023.

The NBU sampled a large number of yellow-legged hornets from Kent but found no clear evidence for or against an established population. It could be that a population has established in Kent, or perhaps a number of mated queens were helped across the Channel by winds from the continent, or the large number of nests might have resulted from a combination of these scenarios.

What will happen in the UK if the yellow-legged hornet establishes and starts to spread? Well, we will look at strategies to eradicate and contain the hornet next but, in addition to several studies that look at spread throughout Europe, including the UK, one study (Keeling et al. 2017) has attempted to model the spread of yellow-legged hornets specifically through the UK based on establishment at Tetbury (where the first UK nest was found). The authors show that, without

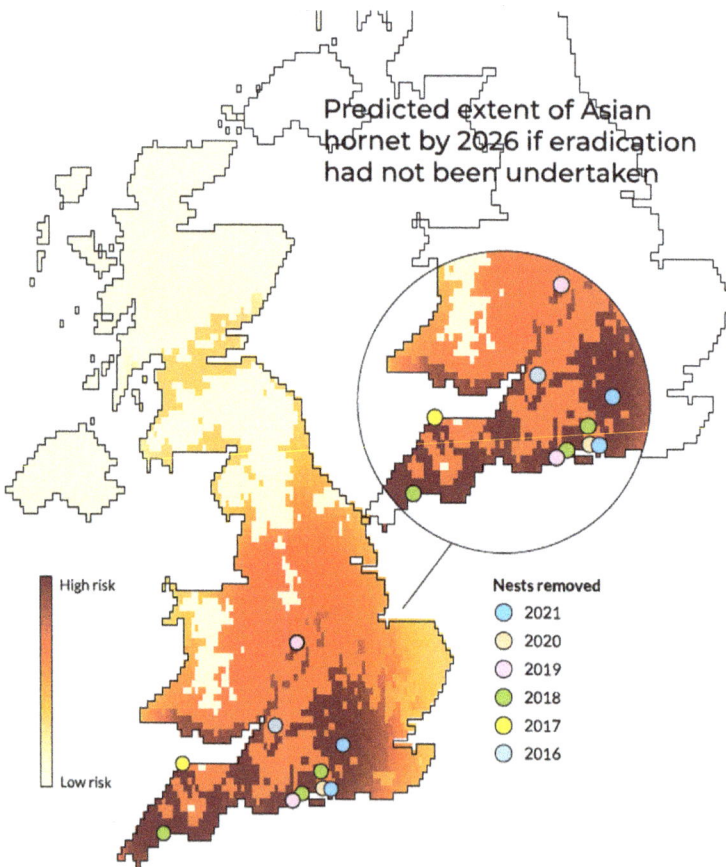

Map of the UK showing the predicted risk of YLH spread by 2026 if eradications had not been undertaken. The locations of YLH nests that were identified and removed between 2016 and 2021 are shown with coloured spots.
Source: UK Centre for Ecology and Hydrology. Crown copyright 2023.

control, it would be possible for *Vespa velutina* to become established across most of England and Wales in 20 years (petering out along a line roughly from Newcastle to Carlisle) — unless our temperate maritime climate is a strong deterrent. This timescale does seem a little optimistic. In this scenario, although moderate detection and destruction of nests would slow the spread, extremely high levels of detection would be needed to eradicate yellow-legged hornets completely.

Another study, by the UK Centre for Ecology and Hydrology, looked at which areas would be at risk of establishment by 2026 if no nests had been destroyed in England, that is over a 10-year period (see opposite). Hassall et al. (2024) are convinced that active eradication programmes by the UK, Germany, the Netherlands and Belgium have led to better outcomes in these countries than would have occurred with no attempt at eradication.

Face of yellow-legged hornet queen. Photo by Chris Issacs.

Part III
Control

Strategies for Control

The yellow-legged hornet is classed within the European Union as an Invasive Alien Species (IAS) of Union Concern, because of its rapid spread through member states and across borders. An IAS is an alien species whose introduction or spread has been found to threaten or have an adverse impact upon biodiversity and related ecosystem services. To try to counter this invasion, member states are obliged to try to eradicate it, and when that fails, to manage it as best they can.

The Australians, who know a thing or two about invasive alien species, reckon that money spent during the early stages of an invasion gives you vastly better economic returns than later, from 1:100 at prevention, to 1:25 during eradication, 1:5–10 during containment, and 1:1–5 during management to protect assets (Victorian Government, 2010).

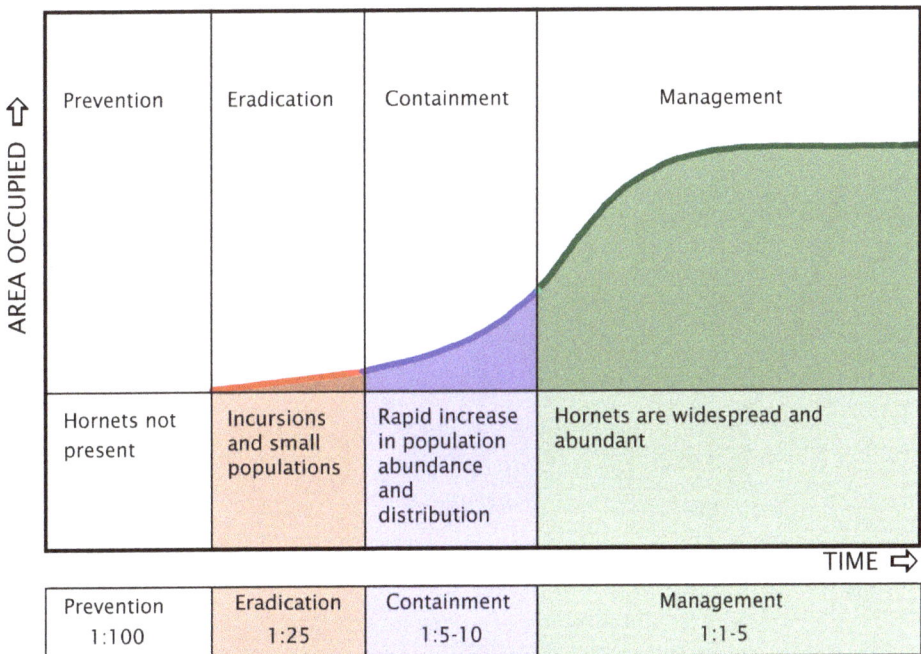

Prevention	Eradication	Containment	Management
1:100	1:25	1:5-10	1:1-5

Economic returns. Spending money on *prevention* is 20 to 100 times more efficient than spending it on *management*.

Generalised invasion curve showing actions at each stage. After Victorian Government publication Invasive Plants and Animals Policy Framework.

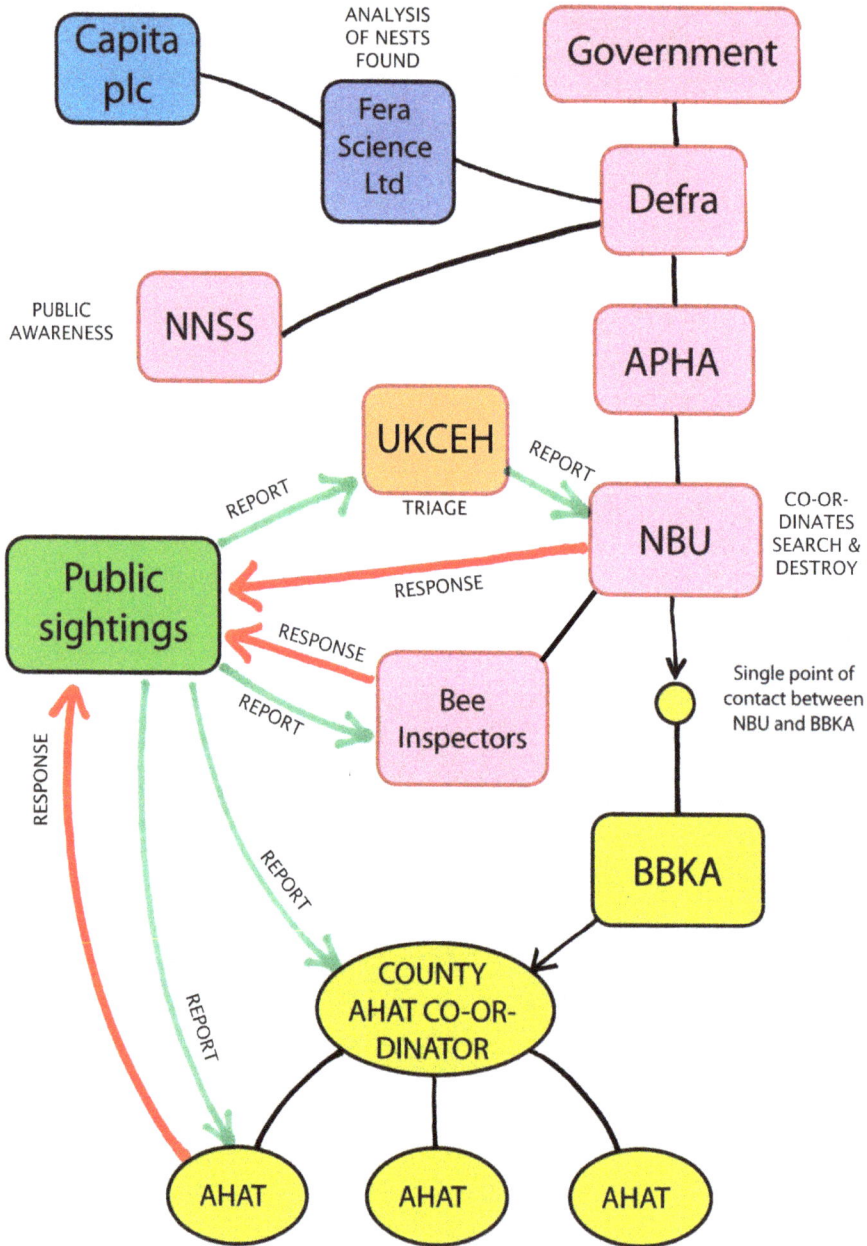

Diagram of main players involved in YLH sightings and responses.
Acronyms:
Defra, Department for Environment, Food & Rural Affairs;
APHA, Animal and Plant Health Agency;
NBU, National Bee Unit;
NNSS, Non-native Species Secretariat;
UKCEH, UK Centre for Ecology and Hydrology;
BBKA, British Beekeepers' Association;
AHAT, Asian Hornet Action Team.

A national contingency plan

As you would hope, the British governments of England, Wales, Northern Ireland and Scotland have contingency plans in place for the eradication of the yellow-legged hornet, and have had time to hone these plans and put some into practice since the first sighting of a hornet in Tetbury, Gloucestershire, in 2016. These contingency plans have been informed by the GB Non-native Species Secretariat (NNSS). The latest Pest Specific Contingency Plan for Asian Hornet (*Vespa velutina nigrithorax*), put together by Defra (the Department for Environment, Food & Rural Affairs) and the Welsh Government can be downloaded from the National Bee Unit website:

https://www.nationalbeeunit.com

The main players

The Pest Specific Contingency Plan for Asian Hornet has two pages of acronyms in the glossary! I have sketched out the main players (diagram) to give an idea of how they are related. Government organisations have their own top-down hierarchy, intended to deal with pest incursions to protect the public, livelihoods and the environment. In the government organisation, the Defra sub-department of the National Bee Unit (NBU) is the main player, and tasked with finding and eliminating nests; surveillance; investigation of single hornet discoveries etc. The UK Centre for Ecology & Hydrology (UKCEH) is an independent charity that collects reports of yellow-legged hornets from the public and passes on credible sightings to the NBU. They also gave us the excellent Asian Hornet Watch app. The GB Non-native Species Secretariat (NNSS) is a government body that helps co-ordinate action on non-native species. They have produced a useful alert poster, an ID sheet and other educational material. The NBU can send you hard copies of these posters, and also cards that are helpful in bringing awareness to the public. The NNSS also contacts all sorts of organisations connected with the natural world, community groups and outdoor leisure to raise awareness.

There also exists a grass roots collection of small teams called AHATs (Asian Hornet Action Teams, pronounced 'ay-hats') that are affiliated with the British Beekeepers' Association (BBKA). AHATs arose out of a desire among beekeepers to have more agency in keeping the UK free of yellow-legged hornets and the idea is that every branch or association under the umbrella of the BBKA has an AHAT, along with a co-ordinator at county level. Although AHATs presently consist of just beekeepers, in future this will need to expand to include other eager volunteers from all walks of life.

There has been some tension between the NBU and AHATs. Some AHAT members (many of whom have trained to track yellow-legged hornets in Jersey) feel that they could assist the NBU to a greater extent than is presently permitted. So far, AHATs have been promoting public awareness, checking possible sightings and have been involved in some NBU tracking. Because *Vespa velutina* is an invasive alien species, a licence is required from Natural England for release of the insect, even when it has just been retained for a few moments to mark.

Waiting…and spotting

In 2011, the Non-native Species Secretariat (NNSS) put together a risk assessment for *Vespa velutina* (Marris et al. 2011). In it, they tried to find out as much as they could about the insect, work out how it could get into the UK, and which paths were most and least likely. They concluded that the arrival of *Vespa velutina* into the UK was very likely, as was its rapid establishment and spread. Less certain at the time was its impact, due to a lack of reliable data from mainland Europe and elsewhere.

The first risk to manage is the arrival of *Vespa velutina* into the UK, but with the amount of traffic between the UK and mainland Europe, and the possibility of hornets also making it across the Channel under their own steam (or helped by strong winds), this remains something that cannot be dealt with. And, looking at the widely distributed sightings so far, the risk assessment report was correct in reckoning that *Vespa velutina* could turn up anywhere, via any one of several pathways.

There is a network of sentinel hives all over the country, and especially near ports and airports, which are managed by beekeepers on behalf of the NBU. They are constantly looking for possible bee pests, such as small hive beetle (*Aethina tumida*), which is expected to turn up at some point. In addition to the beekeepers with the sentinel hives, everyone else in the country can act as the eyes of the NBU.

Beekeepers are probably the most informed about yellow-legged hornets at the moment, because of the threat to honey bees. There have been quite a few articles in beekeeping magazines, and Stephen Martin's book *The Asian Hornet: Threats, Biology & Expansion* (2017). But it is important to emphasise that this isn't just a beekeeping problem: *Vespa velutina* is a voracious predator of other pollinators such as solitary bees and flies, as well as many other small creatures, including wasps, caterpillars and spiders.

Monitoring is vital if eradication is to be achieved, and this is something that the NBU has always had in place when removing nests. After the 2023 season, monitoring networks with one trap per square kilometre were set up by the NBU within a 5-km radius of nests found the previous season. Beekeepers were also encouraged to monitor with traps within these zones, using the most selective trap possible (e.g. a mesh trap, see 'Trapping'), and emptying by-catch daily to reduce the impact of trapping on native insects. The BBKA put out monitoring guidance in the form of traffic-light zones (see p 160).

In 2022, the UKCEH received around 5003 possible sightings of yellow-legged hornets, and 4 of these were confirmed as actual YLHs. In 2023, encouraged by publicity surrounding the large number of nests found, the possible sightings arriving at the UKCEH went up to 20,713, with 150 being confirmed as YLHs. At present, this triage is done by people rather than AI. If the insect is not a yellow-legged hornet, but identifiable, a reply is sent back to the reporter, telling them what the insect is. A positive identification or a credible sighting of a yellow-legged hornet gets sent immediately to the NBU, triggering their response. The problem is the large number of unidentifiable insects. For these, an email goes back to each reporter with links to further identification sites, and encouragement to get back with a photo/a better photo, or a specimen.

Dealing with an isolated incident

When the NBU receives a positive or credible sighting of a yellow-legged hornet, a local Bee Inspector is dispatched to see whether live hornets can be located. If they are there, then part or all of the team moves to the area, initially setting up widespread surveillance (20-km radius, with a 2-km radius prioritised) and then moving on to track insects back to their nest using a method similar to that used in Jersey (there is also the possibility of using radio tracking: see 'Radio tracking'). Beekeepers registered with BeeBase are notified within a 20-km radius, and those within 5 km are supplied with traps. Local beekeeping association(s) are informed, as are AHATs. Once a nest has been located, the colony is destroyed using insecticide, and the nest is removed, if possible, the following day for analysis.

Surveillance then continues, to make sure that there was only one nest. If workers are still found after a few days, then the search for another nest is initiated. Sometimes it is already clear that there are several nests, from workers leaving a monitoring (bait) station in different directions. If there is any possibility of sexuals having left the nest, surveillance is continued until the following autumn.

The Brockenhurst nest. The YLH fitted with a radio-tag approaches the nest: inset shows the tag (yellow) hanging under but close to the hornet; the fine wire aerial trails behind. Note that the nest appears to be damaged: normally all the comb is protected by the paper envelope of the nest, and the small entrance hole is half-way up the side in a secondary nest like this. Photo by Peter Kennedy.

Some sightings reported are of individuals that have hitched rides and are not part of a live colony: early in the year they may be hibernating queens that have ended up in the UK. These sorts of reports are dead-ends; unable to find any more hornets, there's not a lot more you can do except to keep checking local monitoring traps. If there are more hornets around, they should find the traps.

Losing control

If a nest goes undetected, and the sexuals emerge, mate, and the new queens disperse, then population growth can be explosive. According to the NBU, one incursion could lead to 20 or 30 nests the following year, but this is probably a worst-case scenario. Up to 2023, the nests found in England were much smaller than those regularly found on the continent or Jersey. Because there is a direct correlation between nest size and population, smaller nests should produce fewer males and gynes, leading to a smaller number of nests the following year than you would get from a large nest.

Eradication is possible up to a point, but depends on track and trace operations to find and destroy nests. This in turn depends on dedicated bands of people sometimes operating under very difficult conditions. Although eradication is expensive, the costs are tiny compared with doing nothing and having to deal with the consequences. The large number of nests found in 2023, some of them late in the season (six destroyed after mid October), meant that some sexuals did emerge before nests were destroyed and not all nests were found. Intensive monitoring in affected areas resulted in some queens being caught in the spring of 2024.

The spring of 2024 was very wet, and insect numbers were generally greatly reduced until late summer. In total, 24 nests were destroyed in 2024; some late ones likely releasing sexuals. At the end of the year, Defra decided that there still wasn't strong evidence of an established population, and that they would carry on with the eradication strategy in 2025.

If numbers of nests become unmanageable, Defra will have to decide when to move to the containment phase, at which point the NBU will give technical advice and training to beekeepers, pest controllers and local authorities. Any change in strategy is decided over the winter. The final phase is management.

The 'new normal'

Worryingly, there are no clear plans for a management scenario. Instead, the contingency plan just says that 'Longer-term management options for dealing with the pest will be considered by the LDG [lead government department] meeting'. In a management scenario, it is not yet clear who will pay for destruction of nests, but presumably local authorities would deal with nests that are deemed to be a particular public nuisance — for example around schools and hospitals. We would all have to learn to live with a potentially dangerous species (accidental disturbance of nests being the most dangerous thing), and if urban areas are colonised, we would be coming into contact with yellow-legged hornets more and more. Beekeepers would have to defend their hives and perhaps deal with nearby nests. This scenario might be here sooner than we hoped. In late 2023, various County Beekeeping Associations were already drawing up contingency plans of their own (for example the Asian Hornet Contingency Plan for Hampshire).

In a review of the success of the UK strategy to tackle the yellow-legged hornet (Kennedy & Osborne, 2023a), the authors suggest several improvements that could be made. More surveillance is critical to getting reports that can be acted on: public awareness needs to ramped up significantly, and automated detectors may also help. Nest finding could be improved by the involvement of

AHATs and the use of radio-tracking. The switch in strategy from containment to long-term management has understandably been kept loose to allow for maximum flexibility of strategy approach, but this is leading to uncertainty and stress among stakeholders, who are asking the following questions:

- Will AHATs be expected to take over?

- Will active nest-finding stop? (in which case must beekeepers rely solely on hive defence?) And what of the pressure on native insects?

- How will pest controllers be trained and policed, and will there be any subsidies for nest destruction?

- What will happen with other stakeholders, such as the emergency services, commercial and recreational horticulture, woodland and forestry workers, food processing and retail, outdoor sports and pastimes?

European strategies

Of course, the situation is completely different on the continent. Without the barrier of the English Channel, country borders cannot stop the advance of hornets: only climate and altitude can do that. I'm not saying that the Channel stops hornets from getting here, but it significantly reduces the numbers getting across, either by transport or under their own steam.

Also, in the UK, we are still in the eradication phase. Action for eradication is very different from what is largely happening on the continent where yellow-legged hornets are widely established, and in many places have been for a significant time. When the hornet is established, you can only manage the situation, concentrating on public health, horticulture and apiculture: with present technology there is no ability to eradicate the hornet from Europe.

In France, a national strategy has been a long time coming, with the Groupements de Défense Sanitaire (GDS), a non-profit association able to negotiate with multiple beekeeping partners, pulling together a plan in 2022. The goals of the French strategy are to reduce pressure on bee colonies, continue monitoring, and improve safety for beekeepers, the public and certain impacted professions.

To do this, they want to develop searches for nests (traditionally, nests have not been actively searched for in France: nests when spotted are dealt with), and techniques to do this, train pest controllers and enforce good nest destruction methods; and promote the use of traps that limit the impact on other insects. Despite the lack of evidence for spring trapping (see chapter on Trapping), they do condone spring trapping for 2 months in the spring, as soon as queens emerge from hibernation, **but limit this to apiaries heavily attacked the previous year**, with a maximum of 10 traps within a radius of 500 m around the apiary. Records are to be kept of queens caught, and by-catch. Autumn trapping against apiary attacks is allowed, using the most selective traps possible, with 4–8 traps per apiary, and the prohibition of bottle or dome traps (even with bait absorbed by sponge). Hive defence is encouraged.

Secondary nests are destroyed by pest controllers using authorised insecticides and should be removed afterwards. Removal rarely happens in practice. Nests can also be bagged and incinerated or frozen. Differing funding approaches throughout the country may affect nest destruction: when destruction remains at the expense of individuals, as few as 2 out of 10 nests may be destroyed (FGDON35, 2025).

In Belgium, there are administrative problems because of the four or five governments. In Flanders, the government paid for nest destruction for 2021–2022 (although funding stopped too early in 2022, while gynes were still emerging). In 2023, however, specific funding stopped and whether nest removal is paid for depends on your province, your local fire brigade, or the city or municipality you live in. In some cases people have to pay for nest destruction for nests on their property (as in many places in France), which of course is not an incentive to do the right thing. Before 2021, nests were doubling each year, but in 2022 there were five times as many nests. Yet the authorities only seem to be listening to those who want to spring trap, despite the lack of evidence for this working, perhaps because it is the easy and cheap option, and it feels ike you are doing something.

In the Netherlands, 622 nests were found in 2023. They have a voluntary co-ordinator, one dedicated tracker (Rob Voesten), one dedicated exterminator and no national strategy. It proved difficult to get the beekeepers on board at first and, certainly to begin with, the vast majority of tracking to find nests was done by the dedicated tracker (who has spent 3 years of his life battling the yellow-legged hornet in the Netherlands and trying to educate people about what the hornet will do to biodiversity) and a small team that has formed around him. The tracker makes a deal with each province for the costs of finding and removing nests, but of course these costs rise annually due to the hornet spreading across the country. Nests are found with radio tracking and destroyed using environmentally friendly methods such as vacuuming and freezing, or by using diatomaceous earth instead of poison. Some beekeepers are also involved with tracking and finding nests: no release permits are required, so once the Jersey method has been learned, they can start. However, local pest controllers then destroy nests using poison, and insufficient safety equipment has resulted in hospital admissions due to stings. Recently, (late 2023), there have been some positive developments within local beekeeping associations, leading to a large group of people now willing to help, but still no clear focus on how to deal with hornets, and a lot of education to provide.

In Italy, a project called LIFE STOP VESPA was initiated in 2015, with a budget of over 2 million euros. Consisting of surveillance, early detection, rapid response, and control measures, the strategy was in place for 4 years, and slowed down and contained the invasion to Liguria and a small portion of Piedmont, as well as eradicating small outbreaks in other parts of Italy. The project then contributed to a national strategy. In terms of public awareness, they produced 10,000 brochures, 33,000 leaflets and a documentary film as well as organisation of events and workshops. The project was staffed by 30 people, employed on a part-time or full-time basis. The objective was to contain the spread of the yellow-legged hornet by active nest finding and destruction, and to create a coalition of trained volunteers among beekeepers to detect and destroy nests. The rate of spread reduced from 18 km per year to about 3 km per year. In the final year, 98% of nests were found in areas already colonised, confirming the containment of the hornet. A harmonic radar tracking device was developed and deployed to find nests. A total of 2,205 nests were detected during the project, with 1,871 of these removed due to the project's strategy. A key part of the strategy was funding, with training, nest detection and nest destruction all paid for. Now, with relatively small amounts of money budgeted by local authorities after the end of the project, it will be interesting to see what happens next.

Spain has a thorough national strategy that includes nest destruction (preferably with removal and freezing, with insecticide as a last resort) and autumn trapping, with limited numbers of traps in apiaries (for example, one trap for up to 10 hives, four with more than 50 hives). The use of non-selective traps is prohibited by law, so their use in apiaries needs to be authorised. The aim is to control and stop expansion of the hornet into new territories, and eradication if possible.

In Portugal, the national strategy mainly consists of nest destruction, facilitated by the Municipal Chambers. Reports of nests come from the public, and active nest finding is not undertaken. Spring trapping is not pursued, due to ineffectiveness and in consideration of by-catch. There is a national system for alerting the authorities to nests, but it seems that beekeepers are left to their own devices when it comes to protecting their apiaries, in terms of destroying nests.

Switzerland is just starting to be invaded from the west, and to a lesser extent from the north, by yellow-legged hornets from France and Germany. CABI (Commonwealth Agricultural Bureaux International) has been commissioned to help with preparations and control of the first arrivals. A monitoring system was set up along the French border in 2017, and climate modelling has been done to work out the most likely places that yellow-legged hornets might settle. Interestingly, trapping of native insects is being done to provide a baseline for biodiversity, against which long-term effects of yellow-legged hornet predation can be measured.

In September 2020, the first hornets were seen attacking hives in Geneva, and the nests were found using radio-tracking. For the moment, this strategy is being carried forward, with more teams in different Cantons being trained in radio-tracking with the aim of efficiently finding and destroying nests to slow the speed of the invasion and, importantly, to reduce the impact of the hornets on native insects. Again, funding is an issue, with some parts of the control scheme being done by volunteers, for example triage of reports.

Germany does not have a unified way of dealing with yellow-legged hornets. Different provinces deal with it separately, which means that funding is also not uniform, and sometimes not available at all. The main approach to dealing with the problem is finding of nests using radio-tracking, and subsequent nest destruction.

Jersey and Guernsey

In the Channel Islands, the yellow-legged hornet arrived on the two largest islands in 2016 (Jersey), and 2017 (Guernsey). Since then, the two islands have had very different fortunes and have developed different strategies for dealing with the hornet.

Jersey is about twice the land area of Guernsey, and about half the distance to mainland France (Jersey to France around 15 miles; Guernsey to France around 31 miles). The distance between the two islands is around 23 miles.

Jersey developed tracking techniques in 2017, and Guernsey began tracking in 2018, based on the old bee-colony-finding technique of triangulation. In 2019, Guernsey decided to initiate co-ordinated island-wide spring trapping (they call it 'Spring queening'), setting 262 traps in a 500-m grid over the entire island, with traps being monitored by committed volunteers who release by-catch daily.

	Jersey		Guernsey	
Year	Queens	Nests	Queens	Nests
2016	0	0	0	0
2017	0	13	0	2
2018	3	53	4	8
2019 GT	69	83	10	3
2020 JT	42	38	3	0
2021	101	63	8	5
2022	55	174	9	4
2023	476	335	33	14

Data for queens captured (Guernsey) or caught (mixture of trapped and caught in open — Jersey) and nests detroyed for Jersey and Guernsey by year. Nests are combined embryo/primary/secondary/'tertiary' (a nest that has been rebuilt by remaining workers after destruction; this is fairly uncommon). In 2019, Guernsey began island-wide spring trapping (GT). In 2020, Jersey began widespread spring trapping (JT), but not with the density of Guernsey spring trapping. Data from island Asian hornet Co-ordinators, Alastair Christie (Jersey) and Francis Russell (Guernsey).

Meanwhile, Jersey started trapping in 2020, again with committed volunteers, but not co-ordinated in a grid, and at nowhere near the same density. In 2023 they put out around 350 traps, and are aiming for 500 in 2024, which would give them a similar density of traps to Guernsey. They have more traps in the northeast of the island where there seem to be more arrivals from France. Interestingly, of the 476 queens caught in Jersey in 2023, 298 were caught in traps and 176 (roughly a third of all queens caught) were just found out and about: of these, 135 were caught and killed by members of the public.

So both islands have strategies that include spring trapping and active nest finding and destruction, but Guernsey seems to have settled into annual spring trapping that is quite thorough, while Jersey has a looser strategy that is responding to increasing numbers of nests found each year, probably fuelled by their proximity to France.

Has Guernsey's tactic paid off? Unfortunately, it's difficult to tell. Certainly, Guernsey is in a much better place in terms of numbers of hornets and nests. But is that simply because they are farther away from established populations of hornets? The numbers of nests were already racing away in Jersey by the time Guernsey decided to start its trapping campaign. Is Jersey so close to the French mainland that whatever it does, more hornets will arrive anyway, so they are more like a region on the mainland than an island in terms of hornet dispersal? It will be interesting to see what happens in Jersey once they have the same density of traps in place as Guernsey, but will they be brave enough to stop trapping if it makes no difference in the long run?

A Guernsey modified Véto-pharma trap, with 6-mm holes in the wall. The sponge has a little bait to attract hornets. Other insects still get caught and are released daily by volunteers. Photo by Francis Russell.

Mallorca, a case of eradication

In Mallorca (a Mediterranean island 176 km from Spain), in October of 2015, a single hornet was found by a beekeeper in an apiary. Once it was confirmed to be a yellow-legged hornet, a working group was immediately set up to co-ordinate the following things: (i) assessment of risks and impacts; (ii) finding out what area was affected; (iii) evaluation of treatment options available and their impact on non-target fauna; (iv) obtaining appropriate funding; (v) communication with stakeholders, including the public; (vi) deciding what management options were achievable (eradication or control); and (vii) detailing the planning methods.

From October 2015 to October 2020, intensive surveys were implemented within an eventual radius of 28 km from the original sighting, by setting traps at 300-m intervals along secondary roads or private paths. Each year, monitoring was carried out within a 5-km radius of nests found the previous year. Meanwhile, a public awareness campaign was carried out using leaflets, brochures and local media. Two 24-hour phone lines were set up and, in 2016, a website and android app were added for reports to be made. Active observations were made by beekeepers and environmental workers in apiaries and natural areas. Nests were found using a triangulation method similar to that developed in Jersey. Nests were removed manually and frozen to kill the colony. The area was checked for a week afterwards and any nest re-builds were taken down.

This resulted in one secondary nest found in 2015, nine secondary nests in 2016, 20 secondary nests in 2017, a single embryo nest in 2018, and no nests found in 2019 or 2020. More than half of the nests (58%) were found because of public reports of adults, leading to nests being found and destroyed.

Year	No. traps deployed	Weeks traps active	No. queens caught
2017	79	11	17
2018	561	12	8
2019	578	13	0
2020	283	9	0

Low numbers of queens were caught in traps during eradication in Mallorca. Data from Leza et al. (2020).

The capture rate was incredibly low for the spring trapping (see Table above). Because the traps were plastic bottle drowning traps, collateral deaths of native insects would have been heavy, which is why the team decided not to deploy them farther than 5 km from previous nests. Better traps are now available.

Summary

National strategies vary, largely depending on whether there is a chance of eradication or not. And this in turn depends on geography. Those nations and provinces that are islands have the best chance of eradication, but only if they are far enough away from established populations. Jersey could have eradicated yellow-legged hornets if not for yearly arrivals flying in from Normandy. England is in a slightly better physical position, being twice the distance from France, but the land area is massive in comparison, and our people are not yet fully aware of the problem. Mallorca is too far for hornets to reach under their own steam, but they still have arrivals due to traffic, goods and shipping. The greater the density of hornets in the established region, the greater the likelihood of hornets being transported across, or crossing the sea. Even if one incursion is dealt with, constant vigilance must be maintained, and every single incursion destroyed. This depends on high enough numbers of educated, attentive citizens maintaining watch and reporting not only nests, but individual hornets (we're not there yet); and fast, effective nest location and destruction, which we have.

The next most fortunate in terms of geography are those countries which, although contiguous with neighbouring states, have narrow corridors of likely entrance. This includes northern Italy where the expansion has occurred through a narrow strip bordered by the sea to the south and mountains to the north. Such situations may achieve eradication, and if not, then perhaps control in a region: but not without constant battle. The strategies of both islands and 'corridor entrance' states depend on enough reports of hornets coming in from the public, or from surveillance traps, to lead to all the nests via active hunting of nests, and swift nest destruction, followed by surveillance.

Life goes on for a colony transferred to a cage for observation.

Finally, those countries whose borders provide no natural barriers to hornets have a dilemma. Should they try to eradicate, or go straight to management? Spain, Portugal and Luxembourg appear to have decided to go for management. Indeed, France, Spain, Portugal and Luxembourg all destroy nests reported to the authorities, but no attempt is made to actively find the nests of hornets seen by the public. Belgium and the Netherlands are still trying to locate and destroy nests, but funding and finding people willing to do it are problems.

Once the hornet has become established, with present know-how, it is impossible to eradicate. Unfortunately, management of hornets where they are established usually consists of two strategies: (1) passive reporting of nests for destruction, either paid for by the state or by landowners (who then have a disincentive to do so); and (2) mass trapping campaigns, for which the strategy documents usually say something along the lines of 'we know that this doesn't work, and that its effects on native insects are terrible, but we're going to do it anyway'.

Kennedy & Osborne (2023a) point out that adequate funding, resulting in timely action, is crucial. In both Belgium and the Netherlands, inadequate funding has resulted in eradication projects left high and dry at a critical point in the season when sexuals were emerging (Belgium: Dominique Soete, pers. comm; the Netherlands: Michelle Smeets pers. comm.).

Pazos et al. (2022) looked at lessons of best management practices in the control of the yellow-legged hornet in Galicia, northern Spain. They investigated the collaboration between experts and non-experts, and found that lack of adequate environmental education hampered control efforts. One of the most common obstacles was differences in perceptions that different stakeholders (e.g., local communities, authorities, experts) had about management strategies or priorities. This can lead to misunderstandings, inefficiency, and even conflict. Better communication between experts and non-experts is critical, via meetings and information campaigns, either local, in the media, or on social media. This can help control misinformation, irrational beliefs, and sensationalist and misleading headlines.

The public need to know how to identify and report yellow-legged hornets, and also about the risk of stings and costs of control. Beekeepers need to know best practice based on evidence, not loud voices on social media.

Although nest removal is considered the best strategy for eradication, and for slowing the spread of the yellow-legged hornet, lack of reports of successful cases can lead to scepticism among the public about the ability of authorities to manage the invasion. On the other hand, frequent celebrations of successes in the local media, as happens in Jersey, engages the public and leads to a sense of ownership of the struggle, as well as encouraging volunteers to join in.

Pazos et al. (2022) note that in the case of eradication, managers need to secure enough funding to thoroughly complete the job, whereas limitation of spread or reduction of density of the yellow-legged hornet must rely on detailed cost-benefit analyses in order to produce control guidelines (although data to base this on are scarce).

Removal of nests for public health reasons can also involve difficult balancing acts: if removal is by emergency services (fire brigades are often the people with the right equipment: ladders and hoists), then how does this impact on other services they normally perform? And does use of emergency services increase the perceived threat of this species? Such management conundrums can only be solved with good communication and carefully considered management plans.

If yellow-legged hornets become established in the UK, it will be very difficult to get rid of them. The longer we can hold them off, the more possibilities might arise from pheromone-based attractants and targeted poisons, possibly allowing their numbers to be kept down long term.

A worker elicits food from a larva. Photo by Chris Isaacs.

A worker finds a speck of sweet fondant on a coin. Photo by Angus Deuchar.

Finding Hornet Nests

Introduction

It is important to reiterate that at the time of writing (spring 2024), it is not certain that yellow-legged hornets have established in England, even though so many nests were found in 2023. The correct procedure if you think you have spotted a yellow-legged hornet is to try to get a photo, or the actual insect, and report it (see back cover). If it is a yellow-legged hornet, the NBU will look for and find the nest, if there is one, and destroy it.

Also at the time of writing, the Animal and Plant Health Agency (APHA) does not allow the release of an 'invasive alien species' (and *Vespa velutina* is classed as that) without a licence. This is understandable in that this proscription is a one-size-fits-all law that would apply equally to any invasive species (plant, animal etc.) and exists to prevent the spread of non-native species. However, in a scenario where a colony of yellow-legged hornets has established and a steady supply of workers are visiting a bait station, releasing a marked hornet in order to find the nest and destroy it would seem to warrant an exception to this rule. Not only are you helping destroy the nuisance species, but you are releasing a worker that cannot found its own colony, and doing it within its existing home range. Also, in a few minutes it will even come back to you.
Consider another situation, however. You find a single, live yellow-legged hornet in the spring which you mark and release. You didn't realise it, but it was a queen, and she doesn't come back: perhaps she was feeding during her migration. There was no nest to find, and now you have lost what would have been valuable evidence of a yellow-legged hornet in that area. She then goes on to found a colony which is undiscovered and leads to the establishment of yellow-legged hornets in the UK. *If only you had caught it and reported it.*

Yellow-legged hornets could establish in the UK at any time, and I want you to be prepared. However, please follow the national guidelines. I have written the following section on tracking as if a licence has been granted, in order to make it easier to read and absorb.

Note: Many **embryo** nests (the small nests built by the queen in the spring and often found attached to man-made structures) are found by curious and informed people. But once the first workers have emerged from the primary nest, there is no reason why these workers cannot be followed back to their nest. Indeed, nests might be easier to locate earlier in the summer because there is evidence that the workers don't fly as far from the nest (foraging range around 300 m for 50% of workers, Sauvard et al. 2018).

The Jersey method

Developed in Jersey by Bob Hogge and a small, enthusiastic team, this technique has been refined since 2017 by the Jersey Asian Hornet Group, along with associates tracking around the island. In Jersey, permission to release insects is obtained from the Jersey States Department of the Environment (Jersey is not part of the UK and not fully in the EU). This simple technique can be learned quickly and can be practised by anyone.

Essential points

Always tell people who you are, what yellow-legged hornets are, and explain what you want to do. If you are from an Asian Hornet Action Team (AHAT) you may have an information leaflet to give to the land owner; at the very least you should carry some yellow-legged hornet identification leaflets. **Get permission from the landowner to go onto their land.** This is essential, and where local knowledge and connections come in very handy. All you have to do is knock on the front door of a property or call out in a farmyard. If you can't raise anyone, go back later, or try a neighbour who might know where the landowner is. If you don't get permission, either through not finding the landowner or them refusing permission, then try and get permission for neighbouring land instead — you should still get times and flight directions, and in rural or urban settings you should be able to carry on tracking.

- Always leave a contact number so that the landowner can get back to you.

- Never leave unattended bait stations where people or animals could get stung.

- Be considerate when parking.

- Label bottles of wasp attractant, and don't leave them lying around — it looks like fizzy pop.

- Make sure that bait stations are removed when you have finished.

- And of course follow the country code if that's where you are hunting.

Getting set up

Starting from the point of getting a confirmed sighting, the first thing to do is set up a bait station (see Box 1). Ideally, do this with another member of your team; it helps to understand the setting as soon as possible. If you know you can return later (in a couple of hours, or even the next morning) then it is ideal to set up a bait station at the site of the first report. Then, by the time you get back, there should be visiting hornets, saving you a lot of time. To speed this process up, you can release bait scent into the air for the hornets to pick up (see photos). The reporter of the hornet may have great local knowledge and help with contacts in order to get permission to go onto land.

As soon as you know where you will be searching, look at a map and get familiar with the area. You can use a map on your phone or tablet, or print one off using a PC and printer. If printing, aim for an area around 500 m east and 500 m west (1 km in total) of the initial bait station. Aerial or satellite maps are incredibly useful because they show actual trees, hedges, buildings and roads.

What you will need

Essential for noting flight direction: bait station (including extra attractant), map, notebook and pen

Essential for timing: queen catcher, 'uni POSCA' marker pens, some kind of timer (stopwatch or smartphone), notebook and pen

Essential for working in a team: walkie-talkies

Essential when locating nest: binoculars, tracking and data apps (explained later). More bait stations, streamer kit

Good to have: yellow builders' bucket for putting dish bait stations on, plastic electric fence pole to hang a wick-pot on, food, drink, light camping stool (if you need to sit while waiting for hornets to return), sun hat

BOX 1. Setting up a bait station

The best place to site a bait station is somewhere fairly open where you can see the hornet flying off for as long as possible. The hornet vanishes against a complex or dark background, especially hedges and trees, but also houses and rooftops. The more sky you can see, the better. Elevated sites are particularly good because the route to the nest will have fewer obstacles and the flight path is likely to be straighter. Although they will often fly along landmarks such as hedges, walls and even roads, they certainly don't do that all the time, and do not have trouble picking out a bait station in the middle of a field as long as they pick up the scent of the attractant. Obviously, keep the bait station away from the public and animals.

The bait station itself should also be elevated — it makes it easier for both humans and hornets to find. It also makes it much easier to observe and catch insects. Cheap yellow builders' buckets from a DIY store are great — you can see them from a long way off and they can be used to carry things. Turn one upside-down and place your bait station on top. There are two types of bait station: dish or wick-pot.

A well set-up dish bait station showing the amount of bait to use. There is one marked hornet: dead wasps have probably been killed by hornets at the station. Photo by Judy Collins.

DISH: A shallow dish (a plastic takeaway dish with lid is ideal because the lid can be put on at the end of the day to stop evaporation and spills). Next some tissue, kitchen roll etc. folded flat in the bottom of the dish, with some attractant on top. The aim is to soak the paper and to have no more than 1 mm of liquid in the bottom of the container. Put a stone or two on the paper, to keep the tray in place and to give a non-sticky landing area for the hornets. *PROS*: Slightly easier to catch hornets, especially if there are several; easy to see what is going on. *CONS*: hornets may get sticky feet, attractant evaporates fast and needs topping up. Needs a surface to sit on.

*A couple of dish-type bait stations made by Chris Isaacs. The one on the **left** uses a plastic mesh for the hornets to walk down until they reach the bait which is also absorbed into a cloth. The triangle is lead, to stop the mesh from being blown away. On the **right** is a deeper version where the cone is surrounded by capillary matting soaked in bait to give a strong scent plume to bring the hornets in fast. The advantage of these types of bait station is that the hornet is less likely to get sticky feet, and is therefore more likely to head straight home from the dish, giving a better direction to follow.*

WICK-POT: A glass or plastic jar with a screw-lid. The lid has a hole in it to allow either a tightly-rolled J-cloth to stick through just 1 or 2 cm; or a fan-folded J-cloth if you do not have a release licence (read main text); the rest of the J-cloth sits in the liquid bait in the jar. If you are using fermenting bait and the cloth is tight you might need a small breather hole to release CO_2 so pressure doesn't build up (very messy!). Recommended to also glue a shallow cap (e.g. clean plastic milk bottle cap) to the top of the lid to allow hornets to refuel rapidly from liquid bait. *PROS*: Easy to carry, saves on bait (less evaporation), can be hung from tree or post. *CONS*: Without liquid bait, hornets can take ages to re-fuel, which really slows down the tracking; if the wick-pot is hanging it can be difficult to get it level enough for the small cap filled with liquid bait.

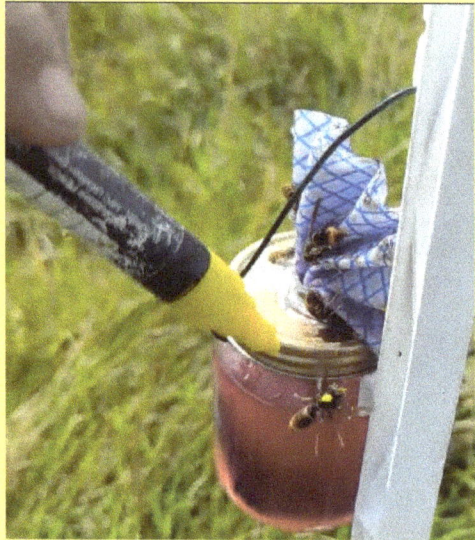

*A couple of wick-pot bait stations. The one on the **left** has a small cap for rapid feeding glued to the lid. The holes for the wire hanger relieve pressure from the fermenting brew. On the **right** is a fan-type wick-pot hanging from an electric fence post (no electricity involved here!). A hornet has been marked without being caught. Photo on right by Sharon Bassey.*

Trappit is a wasp attractant that works extremely well for yellow-legged hornets (Andermatt make a cheaper version). But it can attract wasps, sometimes dozens of them (sometimes none), so you need to be OK with a lot of wasp activity around you. Both the wasps and the hornets are very focused on feeding, but you need to watch out for the wasps when you get your sandwich out! The perfumes in sunscreen and other products can also be attractive to wasps, which can be distracting. Trappit has the advantages of being incredibly effective, and not going off: however, it is extremely expensive.

Because of the cost of Trappit, a lot more people are using home-made baits in Jersey now, and they can be just as effective, or even more effective than using Trappit. The main ingredients are sugar, water, fruit and yeast. The yeast (either wine or bakers', the latter is a lot cheaper) starts the fermentation process, giving off fruity yeasty smells that hornets find hard to resist, and at the same time alcohol is produced that puts off honey bees. **The main thing to look out for when making these concoctions is the CO_2 that is released: any container must allow the CO_2 to escape, otherwise you have a messy, sticky situation on your hands at best, or at worst an explosive situation that could cause serious injury.**

Two bait recipes from Jersey:

1 kg sugar
2 kg water (i.e. 2 litres water, ed.)
Dissolve the sugar in the water. Add 50 g (approximately 2 heaped dessert spoons) of cheap bramble jelly. You can keep this mixture until needed. When you want to use it, add half a teaspoon of baker's yeast. In 10 minutes, it should start to froth up, becoming smelly and attractive.

Chris Isaacs

A weak sugar syrup, quarter to half sugar to water (i.e. 4oz–8oz sugar to 1 pint of water), plus 1–2 tablespoons of 'dirty honey' (honey not good enough to eat from beekeeping processes such as extraction), all brought to the boil, let cool and sprinkle with yeast. Leave to ferment, perhaps making a gallon and fermenting it in a demijohn with an airlock.

Bob Hogge

Below, a yellow-legged hornet feeds at a wick-pot bait station. The liquid bait is pulled up from the jar by the tightly-rolled J-cloth. Over time the cloth can be nibbled away. Notice the hole in the lid near the wick to relieve pressure. Photo: Helen Tworkowski.

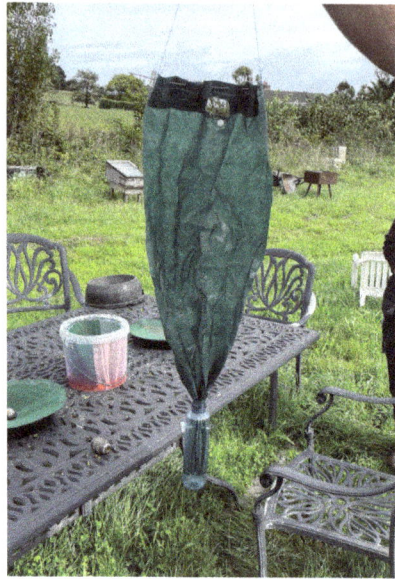

*Methods of getting bait scent into the air to attract hornets quickly. On the **left** the author is using a spray bottle. On the **right** is fabric which can be sprayed with bait and hung from a tree. Excess bait trickles into bottle. Developed by Chris Isaacs. Photos by Angus Deuchar. Pollenize have also developed an electronic mister to do this job (www.pollenize.org.uk).*

Catch your worker

A team of six people is ideal, then you can work with two people at each of three bait stations and communicate by walkie-talkie.

Immediately you arrive, you can look for the flight direction of hornets that are taking off from the bait. You can use a compass; otherwise note carefully the landmark that best shows where you saw it disappear. A distant landmark will give you a more reliable direction to record than a closer one. Trees and buildings are the most obvious landmarks.

Stand well back from the bait station unless you are actually catching, or you may obscure the hornet's view and change the landmarks she is looking for. Leave any hornets alone until they have settled and are feeding; let them feed for half a minute before you try to catch one. To get an idea of flight direction, any hornet will do, but to time a hornet you will need to mark it!

I know — after what you have read about these insects, the thought of catching one without a 6-mm-thick full-body hornet suit sounds scary but, in reality, when they are a long way from the nest and totally absorbed in re-fuelling, they are quite docile. You do not need protective clothing against hornets when tracking, but you might need protection from nettles, brambles and ticks, including sturdy footwear. A piece of beekeeping equipment called a queen catcher is the easiest thing to use to catch your hornet (see photo), the plunger-type is simple to use; you can get them from beekeeping suppliers. Move up to the bait station gently and slowly and carefully place the tube right over the hornet, making sure you don't trap her legs, and wait for her to start walking up the inside. You can encourage her to start climbing by gently moving the wall of the tube against her head. You don't need to catch the hornet in a net

Yellow-legged hornet ready for marking in a queen-catcher that has been modified to make a slot: it can be used for marking and for attaching a streamer.
Photo by Angus Deuchar.

first: it will make her angry and there is a chance of getting stung. Don't chase the hornet around with the tube like I did on my first attempt: you'll just spook her.

Keeping the tube vertical (with the opening at the bottom) to take advantage of their natural desire to move upwards, lift the tube and

Marking an yellow-legged hornet using a queen-catcher and uni POSCA marker pen. Photo by Sue Baxter.

swiftly pop the plunger into place, moving it gently up the tube until you have the worker pinned against the grid at the top with her top side ready to mark. If she's the wrong way up you will need to back the plunger off and let her move around until she is standing on the foam of the plunger and try again, using just enough pressure to hold her so that she can't wriggle around. Some people snip out some of the plastic struts that the grid is made from to leave a slot for marking (but still small enough that she can't climb out). Now you can use the 'uni POSCA' markers through the hole or slot to gently mark the thorax (between the wing bases) and perhaps the abdomen. As soon as you have marked her, lower the plunger a little so that she can move and the mark can dry. After 30 seconds or so (to allow the ink to dry), lower the plunger and re-introduce her gently to the bait so that she can have her fill and leave when she is ready. I have seen one get aggressive with the plunger, ripping off good-sized chunks of the foam, but she didn't remain aggressive when replaced gently onto the bait station. There have been very rare cases of a hornet fainting when being marked. If this happens, gently tip her onto a dry surface to recover. See Box 2 for information on colour coding.

If you need to mark a worker without catching her first, let her feed for at least 30 seconds and then move in very slowly and gently with liquid Tippex on a foam or brush type applicator. One or two dabs to the thorax or abdomen should give you enough of a mark to distinguish that individual. Avoid getting Tippex or ink on the hornet's head, eyes, wings etc. A useful technique recently developed in Kent is to use a fan-type wick-pot (see Box 1). The hornet sticks her head between the folds of J-cloth in order to feed, which act like blinkers on a horse. Therefore she doesn't see the pen or brush coming up and can be marked more easily.

Stand back for a good view, make sure your stopwatch is zeroed, or your timer is easy to see, and you are ready for take-off.

Off she goes!

Experienced trackers begin to get a feel for when a hornet is about to leave. She is less focused on feeding and may start to clean her antennae or legs. Some appear to give no warning. There she goes! Hit the stopwatch button or note the time on the timer to the nearest 10 seconds (exact time measurements aren't needed and this will make the maths easier when she returns).

Getting a visual fix for direction of flight is so much easier with two people. Although it may seem pantomime-like, the two of you pointing at the hornet with outstretched arms, and giving an out-loud commentary on its position can really help. She may make a

BOX 2. Colour coding your hornets

When you are working as a group, tracking in one area, or two groups are working together, it is useful if each tracker/pair of trackers have their **own colour for marking the thorax** of all the hornets they catch. After marking the first, add a colour (any colour: the paler ones show up best against the black of the hornet) to the abdomen of the next, so the hornets can be recognised individually. For each, record the colouring, place, date and time.

When you move your bait station towards the nest, you can see whether they continue to visit, or whether you start intercepting hornets from another bait station. Experience suggests that workers choose the nearest bait they can find and abandon more distant baits.

Knowing where you first marked a hornet can help you/the team to understand the range and flight-lines of workers from a nest. One blue thorax and pink abdomen worker faithfully returned to a moving bait station over 3 days. The pictured hornets were probably gathering soft bark to extend their nest, which the blue and pink worker led us to find, 610 m away.

Photo and text by Sue Baxter

couple of orientating circles above the bait, then very swiftly she will head off, often becoming quickly lost against a background, or disappearing over a house.

Where there is a clear line of sight over a good distance, you might be able to watch the flight itself with binoculars. In open fields you can follow a hornet against the sky until she is too far to follow. Using binoculars for this needs a bit of practice: stand a short distance back from the bait, facing the direction of flight. Focus the binoculars on the hornet on the bait and wait for her to fly. Here's the tricky bit: the first few seconds of flight needs very rapid adjustment of your binoculars' focus, so be ready with your finger on the focus wheel and pull it round fast while following the hornet. If you can keep it in sight for those early seconds it is quite easy to follow after that, because the focus changes only gradually once it has reached a certain distance away.

Fix the point of disappearance in your mind. A notch in the canopy of trees? To the left of a distant block of flats? Mark it on your map. Take a bearing if you know how to use a compass. Then wait for her to come back, staying focused on the dish so you don't miss her arrival.

When she gets back, stop the stopwatch or note the time (to the nearest 10 seconds).

We're going to use a rough estimate of every minute spent away from the bait station equating to the nest being 100 m away. This rule of thumb was developed by John de Carteret who, by carefully plotting bait stations and nest sites on Google Maps, then comparing times given to him by trackers, noticed that this was a pretty reliable figure to use.

So if your first time is 5 minutes 37 seconds = roughly 5.5 minutes = roughly 550 m to the nest. A more accurate distance can be calculated by taking the time spent at the nest into account using this formula: (return time in seconds -36) × 1.72 = distance to nest in metres. So, for the above example, 5 minutes 37 seconds = 337 seconds, and (337-36) × 1.72 = 518 m to nest.

Don't just go with the first time you record, but don't hang around getting a dozen flight times. Two or three should suffice. Out of your flight times, the quickest time should be the most accurate (see Box 3). Mark two or three hornets if you can, and time them. It might help for one person to focus on one or two hornets to get a feel for their behaviour. Are they all heading off in the same direction? If not, you may have more than one nest around. Are they coming back with

BOX 3. Flight times and directions — confounding factors

Various things can slow hornets down as they travel to and from the nest:

Complex route — YLHs seem to navigate by using landmarks, so may not take a direct route. They have been seen flying along hedges, walls, roads and valleys, and then turning corners to another landmark or the nest. Obviously this behaviour will affect time and direction. With repeated visits to the bait, they may develop a more direct route as they learn the way. The return times will then reduce and give a more accurate estimate of the distance to the nest.

Resting/cleaning — Return times are also much delayed by hornets pausing after leaving the bait station to spend time on a nearby branch before setting out for the nest. This can occur when their feet are sticky and need to be cleaned, particularly when they have walked up and down on sticky paper trying to find wet bait. Bait stations such as wick-pots are less likely to cause this problem. Jersey trackers think this is crucial for getting accurate travel times.

Wind — Depending on strength and direction, this can slow or speed up different sections of the journey.

Obstacles — They tend to go around or over obstacles, making detours that will make their journey longer. Sometimes they need to ascend a steep slope by flying along invisible switchbacks.

Time at nest — At the nest they offload their gathered bait before heading back for more. By plotting distance to nest against flight time, the Jersey Asian Hornet Group discovered an average unloading time of 36 seconds. The closer to the nest you get, the more significant this unloading time becomes as a proportion of return flight time.

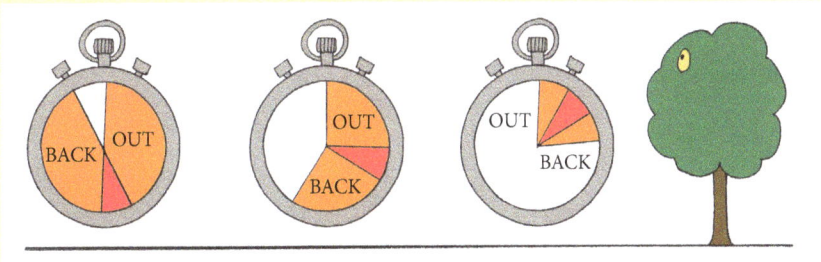

As you get closer to the nest, the hornet spends less time flying to and fro (orange) and the same amount of time unloading at the nest (red)

similar times? Direction is generally a better indicator of multiple nests than different times. Which do you feel is the most reliable hornet in terms of direction and timing? This is the one you should follow.

If there are only a few hornets visiting the bait station, it might mean that it's not in a very good location, or it might mean that the nest is very small and only has a handful of foragers. It's still worth tracking, even with only one or two marked hornets. On the other hand, you might have many hornets and they are leaving the bait station in three or four different directions. This points to multiple nests. If this is the case, you need to concentrate on one of those directions. Choose the most reliable hornet (arrives like clockwork, comes back fairly quickly, always leaves in exactly the same direction), and go after her nest. It's great if you have more than one hornet heading off exactly the same way, but don't despair if not.

Keep on tracking

Direction is more important than time, so don't get hung up on timings. It is useful to have timings because they give you an idea of how far away the nest is (and whether you are getting closer to it!), but it is possible to find a nest by just following the hornet. Keep moving. Once you have some times and a consistent direction, it's time to move your bait station in the direction of the nest, close to where you saw the hornet disappear — but again, try not to get hemmed in. For example, if the hornet is disappearing over a hedge, and the land is open on the other side, then station someone on the other side and see if you can spot the hornet coming over when your partner shouts that the hornet has left the bait station (or use a walkie-talkie). If the hornet is still on track on the other side of the obstacle, then put the bait station on the other side where you have more of a chance of seeing the flight path.

When the bait station is moved, it can take a while for the hornets to find it again. Sometimes they find it straight away, and sometimes they go back to the old location and then follow the scent to the new position. This is why it is important to keep going 'down the line' i.e., following dependable hornets along their flight paths back to the nest (and even moving them in that direction, see next Section).

Repeat the process and keep closing in. Don't succumb to 'tracking paralysis', waiting around a bait station for hours. Try a different tack. Set up some more bait stations. In a straightforward case you can continue until you get the time down to a minute or less, when you can start to search for the nest.

BOX 4. Too close to the nest

"I got it down to a consistent minute and now I've moved the bait station in the same direction I'm getting three minutes! What is going on?"

One situation noticed in Jersey was the timings going haywire if you are too close to a nest. This did not happen for a low nest, but for nests high up in trees.

Suppose you move your bait station in the direction of flight, but you move it to a place directly under the tree with the nest in. You don't know it yet, but the nest is 21 m above your head! Hornets can't climb vertically carrying a heavy load of bait – they will fly at a shallower angle and have a longer flight time. Also, because their normal route out from the tree is at canopy level, they may not recognise the area under the tree. Therefore, it can take a hornet longer to get to and from the nest when it is directly under it, than it did 100 m back along the route you have just taken.

The bait station farther from the tree (blue) gives a more direct and faster route for the YLH than the bait station beneath the tree (red).

3.
Hornets look like they are heading into dense woodland in valley.

2 min 14 sec

1.
YLH discovered in garden centre. Bait station set up.

4 min 37 sec

2.
Permission gained for next field. Unfortunately it had a crop in, so bait station had to be set up at edge, in gateway.

3 min 41 sec

A typical YLH tracking experience, using the Jersey method. Nests are usually located in hours or days, depending on difficulty of terrain, access, numbers of hornets available, weather, trackers available, how far you have to track overall, and whether the situation is complicated by there being more than one nest in the vicinity. Times in the black boxes are the return times for hornets between bait station and nest.

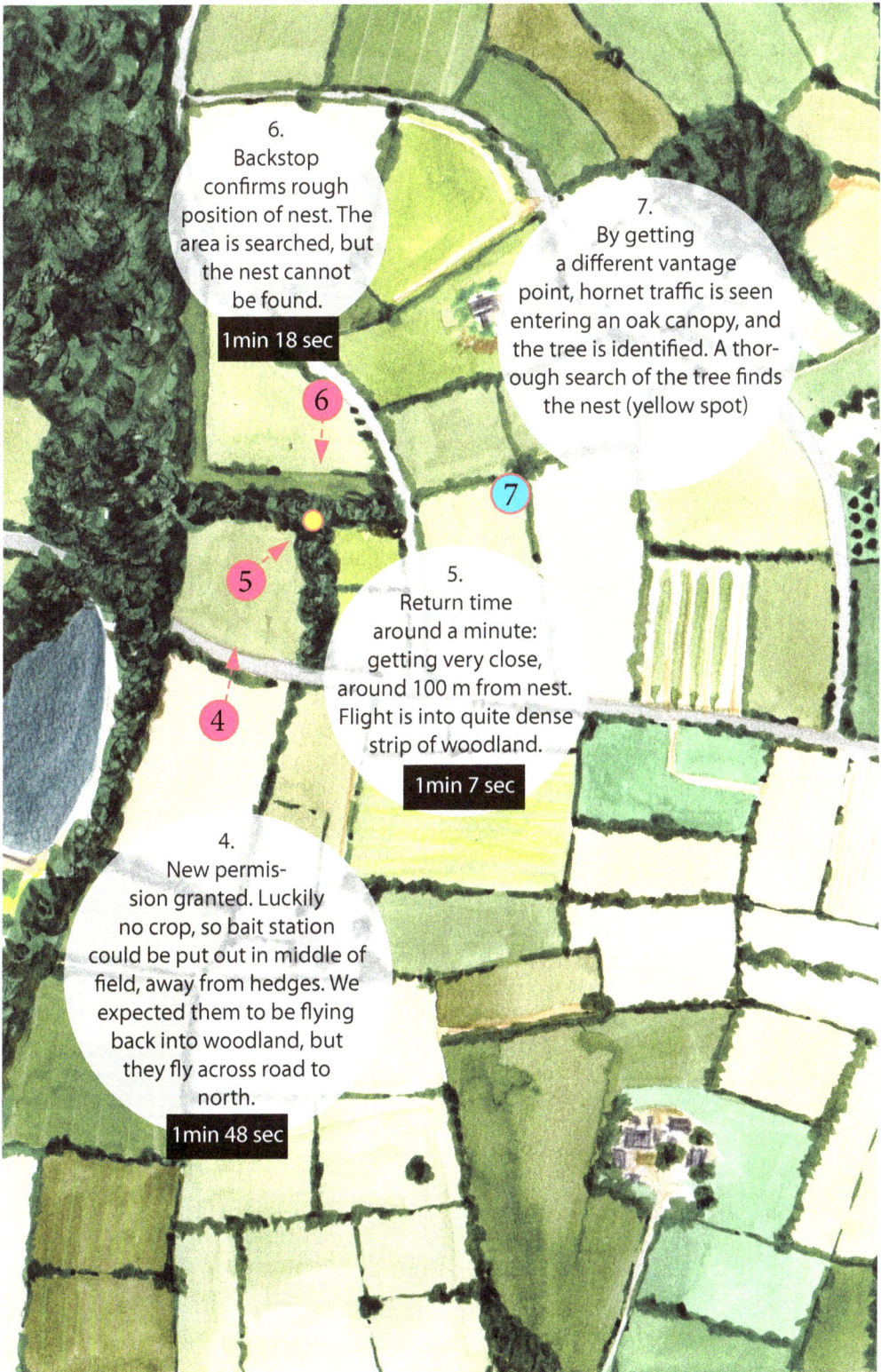

6.
Backstop confirms rough position of nest. The area is searched, but the nest cannot be found.

1min 18 sec

7.
By getting a different vantage point, hornet traffic is seen entering an oak canopy, and the tree is identified. A thorough search of the tree finds the nest (yellow spot)

5.
Return time around a minute: getting very close, around 100 m from nest. Flight is into quite dense strip of woodland.

1min 7 sec

4.
New permission granted. Luckily no crop, so bait station could be put out in middle of field, away from hedges. We expected them to be flying back into woodland, but they fly across road to north.

1min 48 sec

More often, though, you will be faced by some kind of obstacle — a row of houses, a wood, a field with animals in — and you will need to work around. If you have a solid idea of how far away the nest is (from consistent "return to bait" timings), your best manoeuvre is to try to triangulate by placing two bait stations around where you think the nest is. The more equilateral the triangle of bait stations, the more accurately you can pinpoint the nest. Make use of any good vantage points. If you can't triangulate, then try to at least put a bait station beyond where you think the nest is (a "back stop"), and make sure the hornets are flying back towards the nest.

Use your map or app to draw lines from these new bait stations in the direction that the insects are flying — you are now closing in on the nest using three intersecting bearings — theoretically, where the lines cross is where the nest is.

Catch and release from a better place

If you get stuck in a position where it is very difficult to see a flight path, you can move a hornet to a more open space to be able to see her flight direction. This can cut down tracking time considerably. Try to move her 'down the line', i.e. along the flight path towards the nest using the following method.

Once a hornet has landed and has settled, gently cover her with a fairly large container (the cup of a Véto-Pharma trap is ideal, or a plastic pint glass with some small holes in it: the idea is that she doesn't really realise she is trapped), pick up the dish and container, holding them together so that the hornet can't escape, and walk gently to an open place where you have a much better view. Put the whole lot down (if your team-mate has brought the bucket, so much the better), make sure the hornet is still feeding and then gently remove the container and back off. Hopefully she will carry on feeding for a bit before she takes off. Because she is in a new position, she will circle a few times when she takes off, trying to get her bearings before flying to the nest. If you are lucky, you will be able to make out the new flight path immediately. Now you can move the bait station straight away along that flight path 50 or 100 m and wait for her to come back for more delicious bait.

Attaching a streamer

The hornet hunters in Asia have long used a technique of attaching a thread to a hornet, with a small feather attached to it, which makes it much more visible as it flies off (Martin 2017). Nigel Errington developed a technique to do this using fly-fishing materials (on Jersey), and this has been simplified by Chris Isaacs to use thin plastic strips (Box 5). This is an incredibly useful technique, and I would say a game-changer. With it you can see hornets flying into woods and down into brambles. Where once a hornet would have completely vanished against a dark background, it can now be followed. However, this method should mainly be used when getting close to the nest because it slows the hornet down, and she may stop to try to remove it. If you have plenty of hornets, this technique can be used earlier.

A variation on the Jersey method

Rojas-Nossa et al. (2022b) use a variation of this method that might be worth a try (see Box 6). The method is divided into two phases:

Phase 1: Set up a protein bait station (e.g. fresh skinless fish or chicken). Mark workers arriving at bait station and record time away and average direction leaving. Use the measurements for the most reliable (consistent) hornet and mark on a map a line that shows the direction the hornet is taking to the nest. Next, mark lines on either side of this central line at 10 or 15 degrees from the central line. Ten degrees is used for hornets taking less than 10 minutes to return, and 15 degrees is used for those taking over 10 minutes. Next, draw concentric circles on the map that correspond to likely maximum and minimum distances calculated using the formulae. This leads to a likely area to find the nest. If the area is too big, or the nest not found, then phase 2 is initiated.

Phase 2: A second bait station using sweet carbohydrates (e.g. Trappit or something home-made) is set up to one side of the likely area, with the aim of getting a good amount of traffic from the nest to the sweet bait. At this point, it might be enough to visually track hornets or use streamers to follow hornets to the nest. If there is still difficulty finding the nest, Phase 1 is repeated, resulting in a much smaller area to search.

Looking for the nest

Many hours can be wasted by starting to look for the nest too soon. You can see a really promising looking clump of trees, or a particularly large tree, and part of you thinks, 'I could short-cut all this slow, patient bait-moving and heroically find the nest, based just on a hunch...'. OK then, try it — I'll see you by the (now moved) bait station in an hour! Honestly, you will probably want to try this, but after a couple of goes and hearing your team buddies on your walkie-talkie patiently discovering that the hornets are flying away round a corner, you will come round to the methodical approach, which wins out every time.

BOX 5. Attaching a streamer
Technique developed by Chris Isaacs

Kit

Queen catcher with modified grid (with a slot) to allow attachment of streamer, or hands-free home-made version (Box 7, p 200)

Streamers made from shiny, reflective plastic to catch the light. Approximately 150 mm long and 2 mm wide. You can use thermal blankets, party banners and holographic PVC. Cut using sharp scissors. It can be cut more cleanly & efficiently by cutting out squares, stacking them up inside a thin plastic folder & cutting off strips with a guillotine.

E 8000, GSE T-8000, or T-7000 impact adhesive.

Beeswax-tipped stick and thin stick to apply glue.

Method

Catch worker as you would to mark. Gently allow hornet to move around until she is aligned in the slot in the queen-catcher grid.

Place a tiny drop of glue (1 mm across) onto the middle of the thorax, using a thin stick to transfer the glue. Make sure no strands of glue get onto the wings.

Pick up streamer at one end with the wax-tipped stick and gently press onto the glue, spreading the glue a little, then pull away for 10–15 seconds before replacing streamer, making sure that it is absolutely aligned with the body and not at an angle: it must not interfere with the wings. Press gently a few times to make sure there is maximum contact with glue. Detach wax-tipped stick by holding streamer down against the rim of the grid and peeling the stick forwards. Check attached well. Trim length of streamer if necessary (I found about 5 cm, 2" was ideal. If not windy, hornets can carry 150-mm-long streamers: experiment). Gently release the hornet, ideally back onto bait. The streamer should easily slide through the slot in the queen-catcher.

Tips

The longer the streamer, the slower the flight. You want a streamer just long enough to be able to see it flashing in the light. You want to impact her flight as little as possible.

Because the streamer slows the hornet down, you don't want to use it if timed distances are essential. This is an ideal technique for finding the nest at the end of the tracking process.

Don't pick a hornet up by the streamer as it forces them to fly in an upward arc so they can flip over and land on your hand.

The light silver mylar is super reflective, but it is easier to chew off.

Streamers are too much of a handicap for the hornet if it is windy.

If you have been using POSCA pens for marking, the streamer will stick on top fine.

If you have more than one hornet streamered you may want to mark the abdomen to distinguish them.

1. Cut 2-mm-wide strips of Mylar.

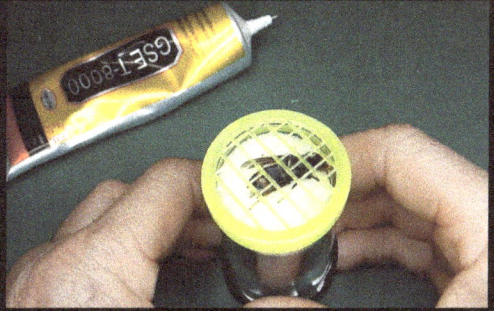
2. Gently position hornet.

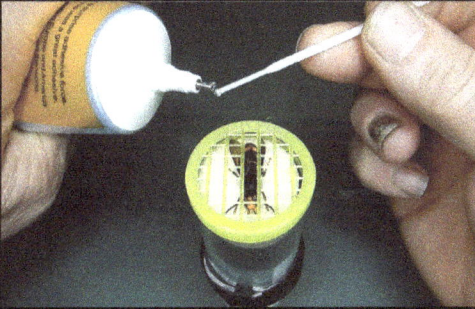
3. Tiny bead of glue onto stick.

4. Glue onto thorax, avoiding wings.

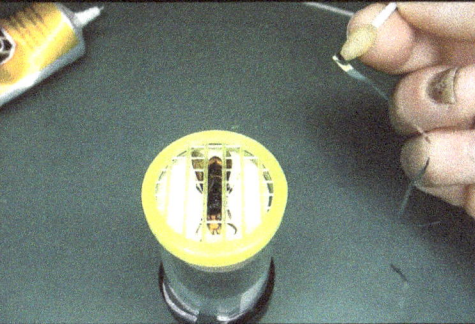
5. Pick up streamer.

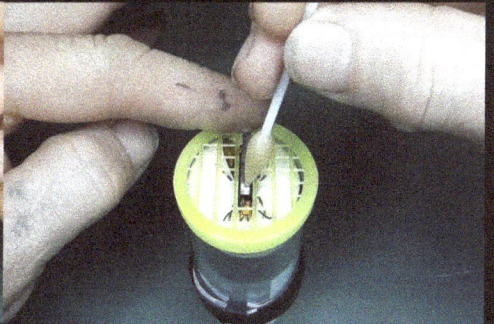
6. Streamer onto thorax.

7. Streamer aligned with abdomen.

8. Release hornet.

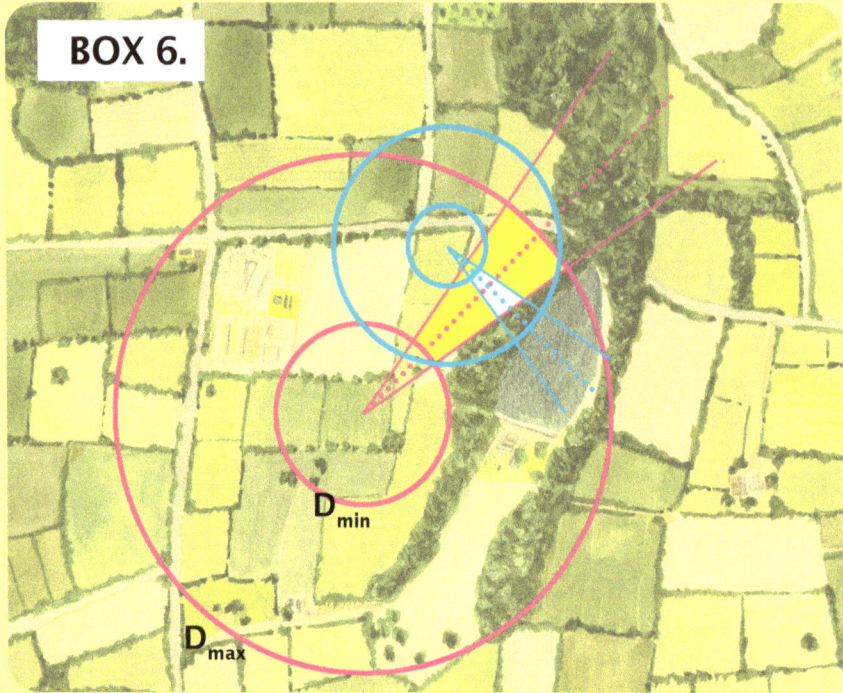

BOX 6.

Method for Nest Detection of the Yellow-Legged Hornet in High Density Areas (Rojas-Nossa et al. 2022b)

Phase 1: Pink. The protein bait station is in the centre of the pink circles. The general direction of departure of the hornet is the pink dotted line. Because the time away from the bait station is less than 10 minutes, lines at an angle of 10 degrees are drawn on either side of the dotted line. The smaller circle gives the distance to the nest if the hornet was slow, and the big circle gives the distance to the nest if the hornet was fast. The nest should be in the **yellow** area.

Phase 2: Blue. A sugary bait station is set up to the side of the area to be searched. The same method is repeated, narrowing the search area to the **white** area. Streamers are useful at this point.

To work out the diameters of the circles, you can use:

$$D_{min} = (t_t - t_n) \div 2 \times S_{min} \quad \text{and} \quad D_{max} = (t_t - t_n) \div 2 \times S_{max}$$

Example: Hornet is away for 5 min (= 300 seconds, this is t_t),
t_n = 45 seconds (always the same: unloading time)
S_{min} = 1.8 metres per second (flight speed from field experiments)
S_{max} = 5.4 metres per second (flight speed from harmonic radar)

$$D_{min} = (300 - 45) \div 2 \times 1.8 = 229.5 \text{ m}$$

$$D_{max} = (300 - 45) \div 2 \times 5.4 = 688.5 \text{ m}$$

It is best to start actively looking for the nest when you are getting return times of around a minute or less, or at least a convincing estimate of nest position from three crossed bearings. Get as many people searching as you can — often a fresh pair of eyes can spot a nest in a tree already searched several times. Nests are notoriously difficult to spot from directly under a tree. It may be better to be on the look-out for 'traffic' (see later) to guide you to a particular tree, and only then spend time looking up at tree tops for the silhouette of a spherical nest.

When searching a tree from the ground, be methodical. If you can (for example with a big oak), follow each branch with your binoculars. Check the tree out from all sides. If you can, try to get a side view of the canopy from a distance, perhaps by climbing up a valley side. When nests are high up in the canopies of big trees, they are often only 2 to 6 feet (60 cm to 2 m) inside the canopy, and a breeze might part the leaves just enough for you to see it. Sometimes the time of day makes a difference — light coming from a different angle can help or hinder.

Apart from picking out that beige ball up in the canopy, nests can also be discovered (or the tree confirmed) by spotting 'traffic' — the fast and direct movement of hornets in and out of the canopy (roof space, hedge, brambles). Through binoculars, tree canopies seem smothered in insects (they are), but most are bumbling around, dipping in and out, flying here and there, and they are all different sizes. The traffic going in and out of a nest is purposeful and unswerving, and the hornets themselves are big and distinctive compared with other insects around. Commonly, the traffic goes into the canopy from above, even vertically above, when it can resemble (with a little imagination) some kind of Roman candle firework.

Most nests are high up in tall trees but some nests are built at much lower levels, in hedges, brambles, wall cavities; a very small percentage are even found underground. Of the 36 secondary nests found in Jersey in 2018, 23 were found in trees (roughly 64%), 8 were found in building 'voids' (e.g. attics or behind soffit boards) or on buildings, 3 were found in hedges and 2 in bramble patches.

Low locations, especially hedges and bramble patches are very dangerous. When the nest is high up, it poses no danger to the hornet hunter, but when it is low down, it can be extremely dangerous. This is because yellow-legged hornets will aggressively defend their nests. **Always be on the lookout for individual hornets** — if you are investigated or bothered by an individual hornet, get out **IMMEDIATELY**. If you accidentally get within a few metres of a nest ('about 5 m', de Haro et al. 2010: so double this at least to be safe), you may be attacked. When this has happened to people, a single sting acts as a signal to stimulate more hornets to attack. People have been hospitalised and have died from multiple stings, and even a single sting can prove fatal if you are allergic or have a condition that makes you susceptible to the effects of the venom (for example some heart conditions). If you are attacked, run away, while trying to shield your face.

UNDER NO CIRCUMSTANCES TRY TO DESTROY THE NEST YOURSELF

Once you have found the nest, make sure you can find it again. Perhaps leave something temporarily at the base of the tree, or use the app what3words, see p 136. Contact whoever is going to destroy the nest (remember that this section portrays a scenario in which licences have been granted, which is likely to be after yellow-legged hornets have established). If the NBU no longer has this role, it may be a local pest controller who will destroy the nest, as happens in France.

Great digital tools for tracking teams

There are two really useful tools for teams to share tracking information and collect data for the county AHAT co-ordinator. These tools have proved very useful in Jersey.

Google My Maps is a free app that can be used on a PC, tablet or smartphone. It allows team members to share real-time tracking data. Using a set of pre-agreed symbols and labels, team members can show positions of an initial sighting, active and inactive bait stations, and information on who did what, where (a label such as 'wick SB 28-Aug' shows that I set up a wick-pot bait station yesterday, so others on the team can ask me whether hornets are visiting). Photos and wordier descriptions, perhaps noting cautions, can be added. Flight lines, either those that just show direction, or those that have been timed, can be added and start to build up a picture of possible nest position. Estimated and actual nest positions can be added (different coloured symbols), and once nests have been destroyed, they can be kept on the map with a different colour again (e.g. grey) to give awareness to later teams. As the picture builds up, it becomes easier for the team to make decisions about how to proceed quickly, and team members can drop in and out without information being lost. Even symbols that show safe and courteous places to park make tracking run more smoothly.

Training to use such a system is best done in person and will be an important task for team leaders if AHATs take on this job at some point.

Epicollect5 is a free app for collecting data that can be easily used in the field on a smartphone or tablet. It is in the form of a questionnaire that a team member fills out, for example when a queen is found in the spring, when a nest is located, or when a hornet is initially sighted. The data collected can include things like where and when a nest is located, how and when the nest is destroyed, and how many people-hours the hunt took. The questionnaire can be set up to just collect data on the things you are interested in, and then the data can be downloaded as a '.csv' (can go into a Microsoft Excel spreadsheet), and can be imported into Google Maps. Epicollect5 can be found at https://five.epicollect.net/. Hampshire Beekeepers Association have built an app called AH Tracker that is based on Epicollect5 and Google Maps.

How long does this method take?

The Jersey method and variations are not necessarily quick. Usually the method will take hours to days. The NBU found that it was taking them usually 2–3 days towards the end of summer, when hornets are plentiful. In terms of people-hours, an average for Jersey is probably around 25, but it can be left off and picked up again when people are available, as long as records are kept, preferably with shareable digital tools like those just mentioned. The main things that slow a search down are bad weather (hornets not flying), small numbers of hornets visiting the bait station (perhaps it is a small nest), difficult terrain, and high density of nests (that can cause confusion unless reliable hornets are found that always head off in the same direction).

Non-hornet tracking dangers

Sometimes the terrain can become impossible if slopes become too steep to traverse safely, or there are cliffs, or vegetation is impenetrable, or the ground becomes boggy. Large doses of common sense are required as to when to stop if it is becoming too difficult. Retreat and talk with the area co-ordinator as to how to proceed, if at all. In England, infrastructure that doesn't exist in Jersey, such as railway lines, motorways, canals, large quarries or other industries will necessitate new ways of operating, where safety is paramount and access much more complex to procure. For example, the NBU had to negotiate access with the railway in 2023.

Working in teams is safer in many ways, from dealing with accidents to dealing with the public.

Zooming in on a nest in a sycamore. Photos by John de Carteret.

Zooming in on a nest in a sweet chestnut. Photos by Judy Collins.

*Secondary nests. **Top left**: nest in hedge — very dangerous and difficult to see. **Top right**: glimpse of nest in a tall tree. Often there are only certain spots where the nest can be seen from the ground. **Bottom left**: nest in sycamore with hornets on the surface. **Bottom right**: a much easier nest to spot, high up in a pine tree, but away from foliage.*
Top right photo by John de Carteret, the rest by Judy Collins.

What Three Words

What Three Words is a phone app for individually naming every 3 square metres on the surface of the planet. If it is difficult to describe a location (of a sighting, a bait station, a nest), you can use the satellite imagery on the app to pinpoint the location, and the app will give you a unique three-word code for that patch. You can then keep that code to record the position of something, or give that code to others and they can find that point again.

Radio tracking

Radio tracking has been used by biologists for many years to locate and track individual animals fitted with radio-tags (transmitters). You are probably already familiar with wildlife programmes in which large mammals, birds or reptiles are fitted with collars or glued-on devices that allow their position to be tracked. This is fairly straightforward when you are dealing with, say, a wolf where the radio-tag weighs a fraction of the tagged individual, but when you want to track insects, you start to rub up against the limits of miniaturisation of the hardware.

The radio-tag generates an individual radio signal, powered by a battery and sent out along a thin aerial, like a whisker. Put simply, the smaller the radio-tag, the smaller the battery, and the weaker and more short-lived is the signal. In order to follow the insect, the operator uses a directional antenna attached to a hand-held receiver, which picks up the signal from the tagged insect. The signal is transformed so that its strength can be seen on a visual display as well as listened to over built-in speakers. Although signal strength is not a reliable indicator of distance, it is an accurate indicator of the direction of the tagged insect.

Prof. Juliet Osborne's team at the University of Exeter's Environment and Sustainability Institute in Penryn, Cornwall, have a long history of studying the foraging behaviour and flight paths of social bees,

Number 32 is fully awake now. She has been gently restrained by a pliable wire which crosses her petiole (waist). The radio transmitter has been attached (black object below abdomen) by tying with cotton thread around her waist; you can see part of the aerial which will trail behind her. Photo by Peter Kennedy.

137

and Defra were interested in how these could be adapted to track hornets. The first problem they met was finding transmitters small enough to be carried by the hornet. Because yellow-legged hornets can carry quite large prey back to their nest, they are at least built to lift and, by experimenting, the team found that a yellow-legged hornet could successfully fly with a tag up to an astonishing 80% of its body weight, including a 10- to 20-cm thin wire aerial which trails behind the tag. There has been some miniaturisation of transmitters since Dr Peter Kennedy started to track hornets, and now a small transmitter (e.g. Lotek NanoPin, ATS T15 Tiny tag) may weigh only around 150 mg, although it has a maximum range of 450 m in open terrain. This means that you need a hornet weighing 188 mg if it can carry 80% of its body weight, or 230 mg for a safer option of carrying 65% of its body weight. This is a lot less than the previous 278–350 mg (depending on the weight of the tag used), making it easier to find a hornet that can do the job.

The tags are attached with cotton thread around the waist, and slung below the hornet. This is a tricky process that is best achieved by cooling the insect in a tube buried under crushed ice until it has become torpid, and then restraining it briefly while the transmitter is tied on. This technique takes some practice, because the hornet begins to 'wake up' in about 2 minutes, and — although dopy — will begin to kick at the thread and tags, making placement difficult. Dr Peter Kennedy has developed an acrylic plate with a restraining loop that holds the hornet in place while you tie the tag on, yet allows the hornet to be easily released with the tag in place.

Recently, other techniques have been developed that restrain the hornet without having to resort to anaesthesia, using a small, clawed hair-clip (Thomas Beissel), double-pronged curl clips (unknown Galician beekeeper) or locking forceps (Chris Isaacs). Practice makes perfect, and if partially wrapped in an insect net, the hornet can be tagged reasonably well. Alas, it is not for the faint-hearted! The advantage of working without anaesthesia is that the hornets recover and perform better on release, especially when using smaller hornets. Being chilled does drain their energy reserves.

Once released, the tag hangs below the hornet but is sufficiently mobile to allow her to fly and walk. Although the procedure for attaching the tag is straightforward, you do need a weighing machine that is reliable down to at least 0.001 g (cheap jeweller's scales, typically around £20, suffice); a good hand-lens or basic dissecting microscope are also recommended — depending on eyesight — to check that knots are secure. After the tag has been attached, the hornet is placed in a small net tent with food so that she can recover after being cooled, and get used to the tag. Once she is showing good signs of recovery, she can be released. This takes about 10 minutes.

Dr Jess Knapp uses the antenna and receiver to track a tagged hornet to its nest. Not all nests are found in the wilds! Photo by Gerry Stuart

The directional antenna superficially looks like an old-fashioned TV aerial, and is excellent at picking up the direction of the radio-tag. This is attached to a receiver, a box that is carried via a neck/shoulder strap. The radio-tag on the hornet produces short 'pips' of radio signal, at regular intervals, that are converted to audible 'pips' by the receiver. Once the hornet has flown off, the operator follows by sweeping the antenna in an arc to determine the direction with the strongest signal, and sets off in pursuit. A clear advantage of this system is that when obstacles are in the way (walls, lakes, private property with no access granted), the operator can move around these and re-acquire the signal from the other side.

In 2017, Dr Kennedy tracked hornets in both the area around INRA Bordeaux-Aquitaine and on the Channel Island of Jersey (in both urban and rural settings). The amount of time taken to actively track a tagged hornet to an unknown nest ranged between 45 and 133 minutes (mean = 92 minutes). Nests were found between 195 m and 1.3 km from release points. The chase can be slowed down by the hornet stopping to rest, feeding on nectar-rich flowers, or attempting to remove the tag. It is not unusual for a tagged hornet to fly to the

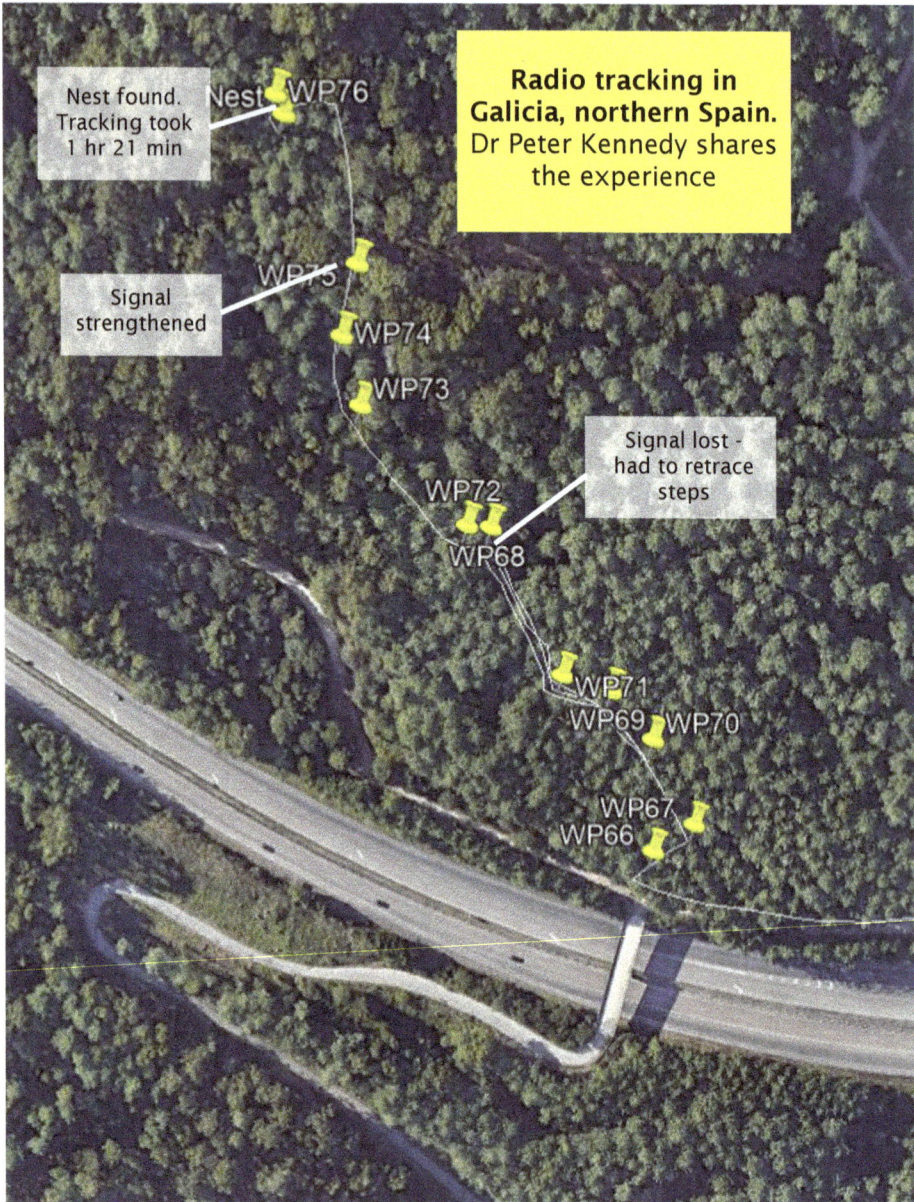

Radio tracking in Galicia, northern Spain.
Dr Peter Kennedy shares the experience

Nest found. Tracking took 1 hr 21 min

Nest WP76

Signal strengthened

WP75

WP74

WP73

Signal lost - had to retrace steps

WP72

WP68

WP71
WP69 WP70

WP67
WP66

A large hornet (0.40g) was fitted with a light-weight ATS T15 Tiny tag (0.17g; i.e. 43% of the hornet's body weight). This was a risk, as such a large hornet would fly quickly with such a lightweight tag and potentially quickly move out of range, but the weather was threatening rain and the tag becoming wet would add to its weight. While the hornet was initially resting in vegetation close to where it was released, I used the opportunity to get ahead of it in the direction we anticipated the nest to be (the other team remained near the hornet). I set off as far I could before the signal weakened (due to an intervening hill) to the point where it was lost and then briefly retraced my steps to where I still had a good signal, and waited. Then things changed quickly as my colleagues initially lost the signal as the hornet flew off quickly and its direction changed rapidly

as the hornet passed me. At this point the heavens opened and we (hornet and trackers) were deluged in torrential rain. There was no stopping the hornet, so I quickly followed up the steep slope to keep in touch with the now weakening signal as the hornet moved farther, and behind a rise in the topography. Cresting the rise, the signal returned strongly, allowing me to circle in on the nest location. A strong signal was detected from the top of a tall pine tree. After a total 1hr 21 min since release (most of which was spent waiting for the hornet to fly from its initial resting place), we spotted a very large, active nest that was 373m from the release point by the bait station. Rain continued and thunder rolled in the distance, while the hornets remained active.

141

nearest tree or bush, when released, and attempt to remove the tag for 15–30 minutes; thereafter tracking can be relatively quick.

Depending on the size and model of radio-tag used, those suitable for insect tracking have ranges between 375 m to 1 km in flat open landscapes and batteries guaranteed for at least 4–12 days. Usually the hornet will return to her nest, so by patiently following her with the antenna, it has proved to be a very successful technique. More recently, hornets have first been trained to a bait station before a suitable one was tagged with a light-weight tag and released. Once the nest is located, the hornet can often be relocated back at the bait station, enabling it to be caught and the tag removed and deactivated for further use, thereby improving the cost-effectiveness of this technique. In practice, the method works best with two people, one person listening to the receiver and the other watching the terrain, getting permissions to go onto new land and keeping a look-out for the nest. Listening to the strength and quality of a signal is easier when other noise, such as that from other excited hornet hunters, can be kept to a minimum. The equipment itself can operate in all conditions hornets do, and although the tags are not cheap (£140 + VAT each at time of writing), they can theoretically be retrieved for re-use.

This technique has been used successfully in France, Spain, Switzerland, Belgium, the Netherlands, Germany, Jersey, Guernsey and the UK, and provides the NBU with an useful alternative when nests have not been found fast enough by triangulation methods. Further miniaturisation of the radio-tags would help extend the usefulness of this approach (Kennedy et al. 2018, Tracking Guide: Kennedy & Osborne, 2023b). A Dutch company called Robor Nature sells a radio tracking kit specifically designed for YLHs, a drone-mounted antenna and a blue-tooth tracking device.

Other methods of finding nests

Using drones (UAVs) and infra-red cameras

Drones, or UAVs (unmanned aerial vehicles), would seem a natural fit to finding big nests in trees that may be hidden by foliage from below. Unfortunately, they are usually hidden by foliage from above, too. OK, so how about fixing up a UAV with an infra-red video camera? As we have seen, yellow-legged hornets heat their nests to around 30 °C. Although nests are well-insulated by paper pockets filled with air, they may still be quite a bit warmer than the background, so perhaps this heat signature could be picked out against a cold background: perhaps at dawn? Well, this technique has been tried, but hasn't been as successful as hoped.

In Jersey, when the very first secondary nest was found, the Jersey States Fire and Rescue Service tested their UAV with infra-red video equipment on board. It was flown close to the nest in the cool of the early morning to see if there was a useful heat signal, but sadly there wasn't. While the drone was near the nest it was investigated by the hornets and collisions with the rotor blades could be heard from the ground. The hornets could be seen attacking the drone, and it was hoped that this unexpected result could be used for nest detection. However, when the drone was used at another nest, no attack response was triggered.

The National Bee Unit tried using a UAV with an infra-red camera at the Brockenhurst nest, but it was unsuccessful, partly because the tree canopy was too warm to provide good discrimination, and possibly because the nest had suffered recent damage and so was unable to maintain its heat.

Lioy et al. (2021b) investigated using thermal imaging to find hornet nests from the ground using a person with a thermal imaging camera. They found that nests could be detected, but the tree canopy was a surprisingly good shield against thermal imaging, and the operator had to get pretty close (less than 14 m) to be able to see nests not blocked by foliage clearly.

A different type of approach was put forward by Reynaud & Guérin-Lassous (2016), who proposed using a small swarm of UAVs to follow a hornet fitted with a lightweight but very visible marker. They assumed that the insect will fly in a straight line back to the nest, which is not the experience of trackers so far (and hornets fitted with markers or tags sometimes stop to rest and/or attempt to remove the object). It will be interesting to see how this research develops, because it does have great potential.

Another way that a UAV could be used is to have one in the air when a hornet has a streamer fitted and is released. It may be possible to follow the hornet with the UAV camera quite a long distance to help find the nest. This is done with *Vespa mandarinia* when a small piece of white plastic has been attached to a hornet. UAVs are regulated in the UK by the Civil Aviation Authority, and heavier UAVs, and those with cameras require a test and flying ID. There are also regulations and approval needed for flying UAVs in urban settings, and of course general rules about flying them elsewhere. There have been studies that have used a UAV to track using radio-tracking, for example Ju & Son (2021), and Kim et al. (2022) who developed an autonomous UAV capable of tracking a (large) insect target in South Korea. However, they still had a tracking error of 20–50 m, which would be quite significant if trying to locate a nest. Presumably, a hand-held antenna could take over at that point to quickly find the actual nest.

Another approach using a UAV, also in South Korea, was that of Jeong et al. (2023), who trained AI to recognise yellow-legged hornet nests, and then mounted this kit and a camera onto a UAV. The UAV was flown on autopilot on a set route, and was able to detect nests in real time. The detection rate was low, although this could possibly be improved with further AI training. The biggest issue was that it could only detect nests in bare trees, after the leaves had fallen. Therefore the method could only be used to find nests that had already produced sexuals, drastically reducing its usefulness in limiting the spread of yellow-legged hornets.

Harmonic radar

Harmonic radars have been used to track insects for many years now. The radar sends out a wave at a fixed frequency and receives a re-flected signal from a metal tag and diode attached to the insect. Pre-viously, this type of tracking has been most successful in flat, open spaces, where the signal was clear; in hilly and woody landscapes it was difficult to make out the signal amongst the noise. In 2016, some Italian scientists (Milanesio et al. 2016) reconfigured the setup, using a small marine radar and advanced processing techniques to pinpoint yellow-legged hornets in particular, in order to find their nests. In 2017, the same team experimented intensely and ended up with a detection range of 150 m. More recently, the same team (Lioy et al. 2021a) has gone further, building a system that can track hornets at a range of around 500 m in flat terrain and 96 ± 62 m (maximum around 300 m) in complex landscapes (meaning woody and hilly). They tend to start from an apiary, where hornets are caught and tagged. If the nest is not within the starting range, the radar has to be packed up and moved closer to the nest. Harmonic radar became instrumental in finding nests during the STOP VESPA campaign in northern Italy, having an effectiveness of 75% at the invasion front, and 60% in high-density established areas. In practical terms, quite a few hornets need to be tagged (35 ± 20), but it only takes a minute or two to tag one. The radar kit is heavy (the old radar system was 50 kg), and has to be taken down and re-set up when moved down the flight path towards the nest. A vehicle is required, and perhaps the whole system could be mounted permanently on an off-road vehi-cle. On average, it takes 2.5 ± 1 days to find a nest using this system (11 ± 4 hours of effective radar use). The system could not detect nests in highly urbanised areas, where there were issues of being able to get the radar into the best position due to private property, and being able to follow the hornet across private land. Wooded areas with few tracks were difficult, as were steep slopes.

Nest Destruction

At the time of writing (spring 2024), the destruction of any yellow-legged hornet nest (primary or secondary) in England should be done by the National Bee Unit (NBU), as part of their eradication programme.

The NBU employs professional pest controllers who have been trained in dealing with yellow-legged hornets, as well as having 14 of their own APHA staff trained in nest destruction. They have specialist equipment: an air lance or cherry-picker to reach high nests, and licensed insecticide. They also remove the nest as soon as possible to prevent insecticides from harming local wildlife.

Hornet suits

Only embryo nests, where there are no workers, should be removed without a hornet suit: an ordinary bee suit is still recommended. Otherwise, a specialist hornet suit is essential.

There are two sorts: traditional ones that are made from heavy, tightly woven fabric, with an integral plastic face-plate; and modern 'breathable' ones made from a special fabric that is meant to maintain that essential thickness of the material so that a sting cannot reach the skin (such as the ultra ventilated bee suit by BBwear). The traditional suits are very sweaty and steamy, especially when you have to start doing physical things in them like climbing trees or sawing through branches; however, they are sting-proof. The ventilated suits are much more comfortable to wear because they are so light and breathable, but they are not always sting-proof. When the fabric is compressed, which can happen when a harness is being worn for tree climbing, a sting can sometimes penetrate through. And the material must not be stretched, for the same reason: they need to be worn baggy to maintain their proper thickness. Jersey hornet workers talked to BBwear and asked them to manufacture an undershirt, made of the same material, that is worn under the hornet suit to give extra protection to the torso and arms. This seems to work very well.

Another problem with the modern suits is the veil, which allows squirting of fluid (see p 73) towards the eyes. As mentioned previously, those dealing with nests in Jersey are now donning plastic visors as eye protection: either strapped over the head or attached to

145

a hat worn inside the suit. Safety goggles or clear skiing goggles are another option. People also wear extra head protection in the form of a hat (e.g. peaked hat) under the suit: head stings are extremely painful (Alastair Christie, pers. comm.). Of course, sturdy footwear and thick gloves are essential.

Destroying embryo nests

In France, destruction of embryo nests (the initial nest containing only the queen and brood, but no adult workers) is often done by beekeepers. Wearing a bee suit, many French beekeepers employ various tactics to remove the entire nest with the queen still inside, including a simple jar-and-card or jar-and-lid technique (AAAFAa 2016)

Equipment needed for nest destruction using a lance (on right, in sections), a duster (at back) and compressed air to propel the insecticide.
Photo by Sue Baxter.

(and finishing them off in the freezer), or a clam-shell device made by taping two small sieves or strainers to the jaws of a litter-picker, for a long-armed approach (AAAFAb 2015). This device can also be used to catch queens on their favourite camellia flowers (AAAFAc 2016). There is more chance of the queen escaping if the outer envelope of the nest has not yet been built, so it might be worth waiting a few days, until the envelope is complete, before removing a nest that you know has a queen inside (it is highly unlikely that she will abscond at this stage).

Once the workers start emerging, a nest should be treated as being as dangerous as a secondary nest, and the professionals called.

Destroying primary and secondary nests

It is important to note that vibration will trigger defensive action from a nest. This can be caused by machinery (e.g. a hedge trimmer), or movement through plant/tree substrates (for example tugging on a bramble stem that is connected directly or indirectly to a nest). Also, in the case of nests in attics or building voids, torch light can trigger defense of the nest, even red light.

Although a nest destroyed at dusk or during the night will result in the whole colony being killed (yellow-legged hornets are generally all home by dark and do not forage again until first light), if the operation does not go according to plan, the hornets can be scattered and end up attracted by nearby house lights. In addition, removing nests at night is difficult and more dangerous for the operators, and also more expensive in terms of paying for work out of normal working hours, so nests are generally treated with insecticide during the day and removed the following day if possible. If a poisoned nest is left in situ, foragers are poisoned on their return. When the nest has been

Destroying a nest in a hedge. The operator is wearing the older-style hornet suit. Photo by Judy Collins.

removed, a trap is left at the site of the nest to catch any foragers. If hornets are still being caught 5 days later (5 good flying days for hornets, NBU method), then it is assumed that there is another nest, and it is tracked. Low nests found in July/August should always be assessed to see whether there is a flight path to a secondary nest of the colony nearby (usually within 100 m).

Another approach is to use beekeeping 'swarm procedure'. A low nest is detached from its support and placed in a big enough bucket so that a tight-fitting lid can be snapped on. The nest is left in the open bucket, as close as possible to its original position, and with the entrance easily accessible. All hornets should be in by the evening, and the lid can then be put on the bucket before taking it away and freezing it.

The insecticide used to kill hornets in their nest in Jersey and England is called Vulcan P5. It is a fast-acting pyrethroid with 0.5% permethrin as the active ingredient (takes 20–30 minutes to kill the colony). Vulcan P5 comes as a fine powder and has to be delivered into the nest as broadly as possible in order to coat as many insects as possible, including the queen. If the nest is low down, then a duster can be used, which uses a pump-action pressuriser. Powder is delivered via a hose and hollow spike, which is jabbed through the nest wall.

Sebastiaan Claus (front) and Rob Voesten remove a nest located in the town of Oegstgeest, South-Holland. You can see hornets on their suits. The nest was located in a black alder tree at a height of 13 m. Drone photo taken by Patrick Obbes.

For nests that are higher, an air-lance is used that is a hollow carbon-fibre pole made up of sections that connect together. Jersey has an AIRadik extendable lance that can reach a nest 30 m up a tree. For nests higher than that, either a combination of cherry-picker and lance is used, or a climber will have to go up and use a short length of air-lance. The insecticide is packed into a chamber near the

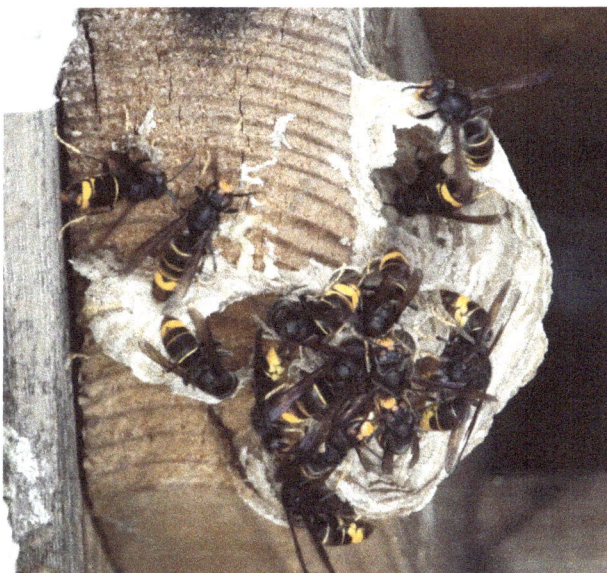

Yellow-legged hornets busy rebuilding a nest that was destroyed during the day: these would have been out foraging at the time the nest was destroyed. Photo by John de Carteret.

top of the lance and compressed air is used to fire the powder into the nest.

Although the pesticide is supposed to remain in the nest, it can blow out of the nest entrance or breaches made by the spike, or the force of the compressed air can blow away the nest wall. It can also be carried at least 100 m by coated insects if the operation is carried out during the day. The NBU and operators in Jersey spread a sheet below nests they destroy to collect any dead hornets which drop out of the nest, and also collect any others they see on the ground.

In France, the lance technique is commonly used, and the same insecticide used. Sulphur dioxide was also used during a short period of time when it was deregulated specifically to deal with yellow-legged hornets (it is still used in some nest destruction in Switzerland). Turchi & Derijard (2018) argue that although sulphur dioxide is dangerous for a user in the event of a leak in an enclosed space, it is much less harmful to the environment than permethrin, and cheaper. Although, ideally, nests treated with insecticides should be removed from the environment, when nest numbers get into the hundreds or thousands, it becomes too expensive to carry on doing that. Nests are rarely removed from high locations such as treetops after poisoning in France or Jersey, but are usually removed from sheds, houses etc.

Another French delivery option for insecticide is a patented modified paintball gun which fires small balls (0.68 inches) into the nest. The balls contain fairly standard insecticides [pyrethroids: the PILP and DIPTER BALL are registered and patented, and you have to have a 'certibiocide' (biocide handling certificate) to be able to buy them]. Paintballs are included in the mix so that the shooter can assess whether they are hitting the target. However, the balls that miss the target will be delivering insecticides into the environment, as well as those that lodge in the nest. Therefore, this technique is meant to be reserved for nests that cannot be dealt with in any other way.

Although some small, low nests in France are dealt with using domestic wasp spray, some sprays have a repellent in them against wasps, so returning foragers are not necessarily killed.

Lee and Yu (2023) describe an attachment for a UAV so that it can be used to destroy yellow-legged hornet nests in South Korea. The UAV itself is pretty chunky: it is an octocopter about 1.5 m long (not including rotors), and weighs 21.5 kg (including batteries). It has a 10-litre pesticide tank, and a flying time of 8 minutes. A 1-metre spray pipe is mounted horizontally at the bottom of the delivery system. Interestingly, they tested delivering the insecticide by spraying the outside of the nest and allowing it to seep through, but found that no spray penetrated as far as a depth of 3 cm, and this approach was abandoned. Therefore, another system was added: a modified airsoft gun for perforating the nest before insecticide was squirted inside. The

Lee and Yu's octocopter in action. The nest is first fired at using biodegradable pellets, then sprayed with pyrethrum. Photo from their article (see references). (CC-BY 4.0). Note that this tree has no leaves.

whole thing was operated remotely by someone who could see video from the UAV, and could use a laser targeter to aim the gun/sprayer (these were designed to coincide at the surface of the nest).

After lab tests, the system was field-tested using pyrethrum extract in the liquid spray, and pellets for the perforator were made from cornstarch. The UAV was manoeuvred into position, first perforating the nest with the cornstarch pellets, then spraying into the nest. Because nests are often built 0.6-2 m down from the top of the tree, inside the canopy, they had to be sprayed from the side, while keeping the UAV out of harm's way (away from swaying branches). In effect, this meant keeping the UAV at a distance of 5 m from the nest. Because the nests were so inaccessible, they were not taken down, but a week later, numbers of hornets entering the nest were assessed, and pyrethrum at 15% and 10% pyrethrum with additives were both shown to have killed the colony. Compared with conventional nest removal (manual nest removal using a net and sprayer, with a vehicle-mounted ladder or extending platform), this system showed a reduction in labour time of 85% and, even though it is a costly piece of kit, they reckoned that the cost of using the UAV exterminator would be 54.9% lower than doing it conventionally, over a 5-year period.

Several other teams around the world have been experimenting with UAV delivery of poison into nests (see video: Chibrac, 2013).

Urgent need for nest destruction that does not require insecticides

There is an urgent need for techniques that will kill off all the adults in a nest, but leave no lasting harmful residue, so that nests can be left in situ. There are already too many pesticides in the environment. Here is what has been achieved so far.

Two non-insecticide options for nest destruction are used in Jersey. The first is a modified Makita hoover (because it runs on lithium-ion batteries and can be used anywhere), similar to those used by bee-keepers to suck up swarms of bees from places with difficult access. John de Carteret modified a hoover so that the hornets are sucked into a separate chamber that can be disconnected without hornets escaping. It is a useful method for low-level nests, such as those in brambles or hedges. Ideally, the hoover is left with the tip of the nozzle close to the nest entrance (achieved by using a clamp or tripod) until all the flying hornets have been hoovered up. Some hornets, including the queen, will remain in the nest. In Jersey, some volunteers are trained in low-level nest destruction, either with insecticide or by using a hoover. Obviously, they are supplied with all the safety equipment they need, including hornet suits.

Another option that has been used in Jersey for nests up to football size is the aerosol pipe-freezer sometimes used by plumbers. The entrance is blocked and then the aerosol nozzle is pushed through the nest wall and the freezing compound sprayed in. When the nest becomes silent, the deed has been done, but hornets may wake up after a bit, so it should be used to knock out hornets while the nest is removed before proper freezing in a freezer for 48 hours.

Because of the problem of insecticides affecting wildlife, other ideas have been put forward for nest destruction, perhaps using fungal spores instead of insecticide (see 'Biological control'), diatomaceous earth, CO_2, water, or steam, and perhaps delivered using a UAV.

Turchi & Derijard (2018) note that other biocides such as pyrethrum or diatomaceous compounds could have less of an impact on the environment than permethrin. Diatomaceous earth is a naturally occurring material consisting of the silicon-based skeletons of tiny creatures called diatoms. It has been used in Belgium, the Netherlands and Jersey. It works by adsorbing the waxes on the surface of the hornet cuticle (skin). These waxes stop the hornet from losing body moisture, and without them the hornet dries out and dies. Used with an air-lance, it has a tendency to clog (Dominique Soete, pers. comm.) and may need re-treatment.

CO_2 has been used in Jersey for high, but otherwise easily accessible embryo nests under roofs of barns, for example (a fairly common place to find an embryo nest). A bucket attached to the top of a pole is filled with CO_2, or has a hose running into it to fill it with CO_2. Because the CO_2 is heavier than air, it just sits in the bucket. The bucket is raised up with the pole to cover the nest so that the nest is dangling down into the CO_2. The queen is knocked out in about 15 seconds, and the bucket can be used to knock the nest off its support.

Using high-pressure water is another technique used to destroy nests. Although the nest is blasted to smithereens, it is difficult to know whether the queen has been destroyed. Late in the season, this probably doesn't matter if the larval and pupal sexuals are destroyed, because there isn't enough time for the queen to build up a whole colony again. With an early nest, there may be time for the queen to start again. However, any traumatised hornets surviving this onslaught could remain disturbed for some time, perhaps causing problems for the local community.

The French pollinator conservation charity, POLLINIS, has been developing a steam destruction method, following how Eastern honey bees, *Apis cerana*, dispatch hornets using a 'heat ball'. Their system, 'HeatNest', is now being further developed by Concept Vapeur, a part of C-Valo, and the product should be available soon as 'ThermoFrelon'.

Ruiz-Christi et al. (2020) also investigated heat death in yellow-legged hornets. They tested individuals of all castes in the laboratory, and found that humid air was actually more fatal than dry air. Steam at a temperature of 92.2 °C killed adults in about 13 seconds. Larvae were more tolerant to heat, but cannot survive for long without adults anyway. The system needs to be field-tested to find out what the response would be when hornets are in the nest: will steam fill the entire nest? Will hornets have enough time to escape? Will the steam kill pupae?

Although flamethrowers (manually operated, or mounted on a drone) are used to destroy some nests in China, setting fire to a nest seems particularly hazardous, whether in an urban or rural setting. Things have gone wrong. Shooting with a shot gun and use of a fire-cracker also seem too risky to be worth it, not only in terms of potential collateral damage, but also because of scattering of hornets.

In Mallorca, Wildlife Rangers with responsibilities for nature reserves remove nests in these sensitive locations without using any insecticides at all. Already trained in tree climbing because of their wildlife duties, they climb up to nests, plug the entrance, free the nest from the tree, bag it and take it down to be destroyed by freezing (Leza et al. 2020). This is a very tricky operation involving a stealthy approach at night, wearing a hornet suit. You have to be extremely dedicated to do that!

Who ya gonna call? In this case, Jersey Asian Hornet Group volunteers, who are about to cut their way into a bramble patch taller than they are, using a battery-powered hedge trimmer and secateurs, find the nest, hoover up the hornets who come out to attack them and release and bag up the nest. The battery-powered hoover is in the foreground, and a post with a kill-trap that will be left at the nest site to mop up any foragers who are presently away.

Using a very tall cherry-picker to remove a nest in an oak tree.
Photo by Gerry Stuart.

Trapping

Context

Insects are in trouble, and we are just waking up to this fact. In the UK, where insects are comparatively well studied, there seem to be the largest documented declines across groups (60% of species). The drivers for worldwide insect decline appear to be (in order): (1) habitat loss, conversion to intensive agriculture and urbanisation, (2) pollution, mainly that by synthetic pesticides, (3) biological factors, including pathogens and introduced species, and (4) climate change (Sánchez-Bayo & Wyckhuys 2019). The authors looked at 73 historical reports of insect decline (mainly in developed countries), and found dramatic rates of decline, which could lead to the extinction of 40% of the world's insect species over the next few decades. This loss of insect biodiversity could be disastrous for the Earth's ecosystems, because insects are indispensable in food webs, are part of many other organisms' life histories (for example, pollination), and help in many great cycles of elements.

Bearing this in mind, we don't want to practice indiscriminate trapping of insects in our single-minded focus of getting rid of yellow-legged hornets. We need to find the best ways of removing the hornets without endangering the others. It's a difficult path to tread, and any action needs careful consideration.

The uses of traps

Traps are used at different times of year for different purposes that relate to the life cycle of the hornet. If you haven't, I strongly urge you to read about the life cycle of this insect at the beginning of this book before reading further.

In places where the yellow-legged hornet is not yet established, monitoring is important in order to find hornets and stop them establishing. Spring queen trapping ('Spring queening' on Guernsey) aims to collect queens that emerge from hibernation to kill them before they go on to build nests (more on whether this works later). Trapping is done in the apiary in summer and early autumn to relieve pressure on hives. Finally, trapping of males and gynes is done later in the autumn to reduce the number of mated queens that then go on to hibernate.

Monitoring/Surveillance

In **spring**, monitoring will only pick up foundress queens (queens that have emerged from hibernation) because these are the only hornets alive at this time of year. Having emerged, (around 12 ˚C: late March, depending on weather) most only within around 250 m of their 'parental' nest, they are intent on building up their reserves by seeking out sugary sources of food: tree sap and nectar. The vast majority of foundresses are likely to build a nest relatively close to the parental nest [95% built within 1.4–6.2 km of previous nests (Lioy et al. 2019)], while a few go much farther. Monitoring traps are useful at this time of year within a 5-km radius of a nest removed late the previous autumn, or in high-risk places like ports. The likelihood of catching a queen in a trap farther than 5 km from her parental nest is very low before YLH are widely established.

In **summer** (mid June to early/mid September), monitoring will only pick up workers. The queen has stopped leaving the nest and is concentrating on egg laying, and the sexuals have not yet been produced. The purpose of monitoring during this period is to find out whether there are workers around, because that must mean there is a nearby nest. This is the time when bait stations can be used to monitor, because workers will keep returning to the bait once they have found it. By using a bait station, you are setting up the first

Summer monitoring is best done using a wick-pot bait station. Photo by Helen Tworkowski.

stage for tracking to find the nest.

Monitoring traps may be used in summer during eradication, to find out where workers are over a large area (e.g., along a country border or within and along the edges of a containment zone) where they can be checked regularly (e.g. by the NBU).

In **autumn** (mid September to end November), monitoring will pick up workers, and sexuals once they emerge. Nests can be found, but once the sexuals have emerged, nest destruction will only kill any remaining sexuals in the nest. However, certainly between mid September and the end of October it will be worth destroying nests, because there is hope that you have got to them before the sexuals have left the nest. Checking the contents of removed nests for dead males and gynes is worthwhile to find out whether sexuals have emerged.

Apart from monitoring traps and bait stations (such as a wick-pot), there are other ways to monitor yellow-legged hornet activity. The first is a simple camera trap that has been designed to take photos of wildlife passing through. A camera trap can be mounted so that it is pointing down at a bait station, and set up to take photos at regular intervals. Yellow-legged hornets can be identified from such photos, but going through the images is time-consuming.

A more sophisticated approach is to use AI to screen for images of YLHs visiting a bait station, and to send an alert when one is spotted. Such a set-up has been designed for use in sensitive areas, such as ports, by Pollenize, based in Plymouth, and another, called VespAI, has been developed by the University of Exeter (O'Shea-Wheller et al., 2024).

Another approach has been made by Jeon et al. (2023) in South Korea, using AI to monitor hives and send an alarm when a hornet is spotted hawking in front of a hive. Herrera et al. (2022) used an optical sensor coupled with AI to distinguish between wing beats of various wasp and hornet species, with the intention of building a monitor for automatic detection of the YLH. Arnia, who produce hive monitoring systems using all kinds of sensors, attempted to devise a yellow-legged hornet alert based on the sound of the hornet, but unfortunately it was not specific enough and was set off by other environmental sounds (such as a chainsaw or pruning work).

Finally, at the University of Georgia in the USA, a project by Lewis Bartlett and others proposes to swab the surfaces of traps or bait stations that have been walked over by hornets, in order to pick up DNA left behind by them. The eDNA (environmental DNA) on the

Jeon et al. 2023 (see References). This camera-based AI system is able to detect yellow-legged hornets hawking in front of hives, and sends an alert to the user. (CC-BY 4.0).

swabs could be amplified and checked against a library to find out whether a yellow-legged hornet had been present.

Can traps of any sort be justified for monitoring YLHs in the UK at the moment?

The NBU strongly encourages monitoring traps, especially along the south coast of England, and provides directions for making them. However, they have moved away from drowning traps and now recommend wick-pot bait stations, a Véto-pharma trap with the bait absorbed into a cloth, or a bottle trap with the bait absorbed into a cloth, and daily release of by-catch. Certainly, there are places that should be monitored, like areas where nests have been found, and in all the sentinel locations, which include ports and airports.

The predicament in the UK at the moment is that we need to know where yellow-legged hornets are so that we can eradicate them; yet, on the other hand, because they have not established, they are rare and patchy in distribution. Monitoring traps can certainly be part of the mix in finding them, but they need to be as selective as possible.

For actions in different zones, see 'Traffic-lights', p 160.If you are a beekeeper, then observe your hives for hawking hornets; if you are a flower-lover, then observe plants that attract a lot of insects, or trees that are weeping sap. We need to monitor, but an unobserved bait station is useless, and a neglected monitoring trap kills insects un-necessarily.

Figure 1. Location of the sampling traps in Asturias (northern Spain) (a); General view of the plot (b); detail of "Vespa Catch" type trap placed on the branch of an apple tree in the study location (c); sample of captured insects in one of the studied traps (d); Vespa velutina specimen from the study site (e). Sánchez & Arias (2021) CC-BY 4.0.

What we are trying to avoid. The authors looked at by-catch from drowning traps in northern Spain, and found 74 other insect groups in the by-catch, with flies being the most prevalent (at 93.26%), but also including Lepidoptera (butterflies and moths) and other hymenopterans (bees, wasps and ants family). The target YLH represented 2.23% (12,835 insects caught in total).

159

RED ZONE

Within a 5-km radius from where a nest has been found in the last year, and any port or place deemed at high risk.

Use selective traps, 1 per 1 km² throughout March-November.

Only trap within an apiary if predation occurs, to avoid attracting hornets.

AMBER ZONE

Within a 5- to 10-km radius from where a nest or hornet has been found in the previous year.

Use bait stations 1 per 1 km² throughout March-November.

Placed in an easily observed location. Selective traps can be used if checked regularly.

GREEN ZONE

Beyond 10 km, and where no hornets or nests have been found in the previous year.

Use bait stations throughout March-November. Place in an easily observed location.

Adjust apiary management to improve protection and promote healthy bees.

British Beekeepers' Association advice in spring 2024 for monitoring.

Spring queen trapping

There have been attempts to catch mated queens in the spring, before they have built an embryo nest, or while they are raising the first brood, to reduce the population in areas where YLHs are established. Such trapping is contentious for three reasons: (1) the amount of by-catch in traps, (2) the small number of yellow-legged hornets caught per trap, and (3) does it even work?

The effect of spring trapping on non-target insects and small numbers of YLHs caught per trap

In 2010, Haxaire & Villemant tested Blot traps in France, plastic bottle traps with separated bait and sophisticated insect escapes (Blot, 2007). They used a bait of lager, cane sugar and rum and tested the traps from the end of March to mid-May. In this experiment, 1,200 insects were caught, mainly ants, and then flies, butterflies (mainly speckled woods) and beetles, various other insects...and eight yellow-legged hornets, in 90 traps over 8 weeks — a catch rate of 0.01 foundresses per trap per week, with YLHs representing 0.67% of the insects caught.

In 2012, Monceau et al. looked at spring trapping, using two types of traps: one a killing trap with no escape holes and liquid bait, and another with a 25-mm entrance and 5-mm and 5.5-mm escapes, with bait absorbed into a sponge (probably the most selective trap at the time). They trapped in two locations, from the end of February to the beginning of May. The overall yield of their trapping was 0.71 foundresses per trap per week, with YLHs representing 1.7% of the total insects and other creatures caught. In their conclusion, they say that foundress trapping is 'a drop in the invasiveness ocean'.

In 2021, Rome et al. worked out that during a season, a colony consumed a mean of about 97,000 prey items (11.31 kg) equivalent in size to honey bees (in actual fact, about 39% of their diet *was* honey bees). Meanwhile, a 1-litre plastic bottle trap can catch around 30,000 non-target insects in a season, and around 20,000 if placed after June and in the vicinity of beehives. *So four to six small traps could catch as many items of prey as a whole colony of yellow-legged hornets,* although the make-up, species-wise would be different.

Many other studies have looked at the low numbers of yellow-legged hornets caught compared with other insects, even where the YLH is fully established. It is important to reiterate that the by-catch in traps is often different to YLH prey, so much of the by-catch *is in addition to* the prey caught by hornets. For example, peak consumption of insects by a hornet colony happens in late summer and autumn, when different insects are on the wing compared with springtime. Unfortunately, unless taken down at night, traps also catch night-flying insects (but not yellow-legged hornets as they don't fly at night).

Does spring trapping reduce the number of yellow-legged hornet nests?

In 2013, Rome et al. surveyed a couple of areas where spring trapping had been conducted by local associations. In the Pays-de-la-Loire region in 2008 a first individual was found south of the Vendée, and in 2009 a limited number of selective beer traps were set up within 30 km of the first sighting. No foundresses were trapped, and 12 nests were later found. In 2010, 400 traps were set up around apiaries, with a big public education effort. Six foundresses were trapped and 195 nests were later found. In 2011, 12 foundresses were caught and 485 nests were found that year.

Elsewhere, a spring trapping operation began in 2007 in the Bordeaux region. It was extremely intensive, with up to 15,000 traps in 550 square kilometres by 2009 (around 27 traps per square kilometre!). Several thousand foundresses were caught each year, but the number of nests found remained stable at about 300 each year.

Similar experiments have been conducted at municipal and departmental scales without any effect on nest densities, comparing years with and without trapping. (Dordogne in 2007, Test-de-Buch 2010, Andernos-les-Bains, 2011, etc.). Conversely, no massive trapping campaign was carried out in the Lot-et-Garonne between 2007 and 2009, yet the numbers of nests detected halved between 2007 and 2008, from 609 to 267, and slightly recovered in 2009.

An interesting view comes from the department of Ille-et-Valaine, the easternmost of the four departments that make up Brittany. They trapped on a large scale in several territories from 2014 to 2018. However, after comparisons were made between numbers of nests found in areas with and without trapping, they concluded that the annual numbers of nests vary in exactly the same way in both trapped and non-trapped areas. This led to trapping being gradually abandoned. The numbers of nests vary from year to year, between around 3,000 and 8,500 (and for the most part moving up and down on alternate years), depending on weather conditions (FGDON35, 2025).

The graph opposite shows numbers of nests destroyed in the four departments of Brittany. All attempt to destroy nests that are found, but funding for destruction may or may not be covered partly or totally by local authorities. Some areas conduct spring trapping; others do not. During spring trapping campaigns, vast numbers of hornet queens may be caught. In Morbihan in 2017, along with 68,264 YLHs, 12,381 European hornets were caught (there is no information on the traps used or whether European hornets were released).

Some of the organisations that gathered the Brittany data opposite (FDGDON du Morbihan and FREDON Bretagne: see 'Resources' for links) joined forces with ITSAP (Institut technique et scientifique de l'abeille et de la pollinisation) and the French Natural History Museum (MNHN) to conduct a 4-year trial of spring trapping, from 2016 to 2019, using data from Morbihan, Vendée and Pyrénées-Atlantique. Unfortunately, at time of going to press, the results have not been published in a scientific journal, and the preliminary results that have been shared (by ANSES, the French Agency for Food,

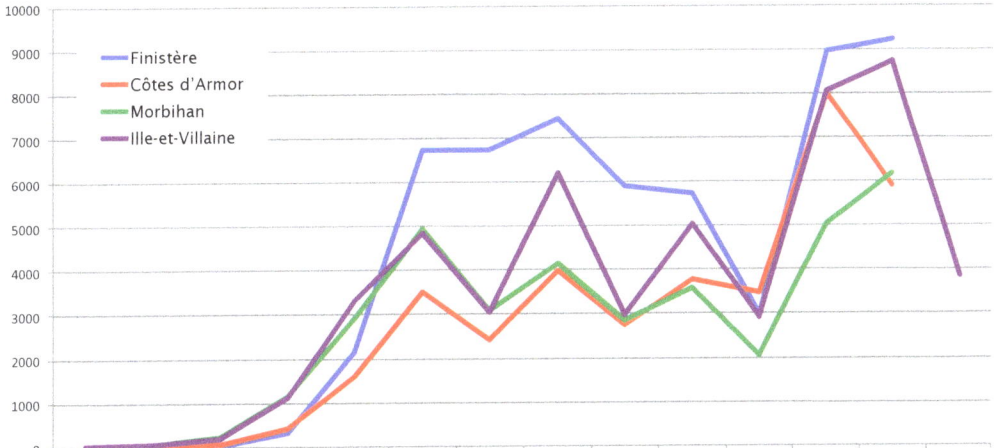

	2011	2012	2013	2014	2015	2016	2017	2018	2019	2020	2021	2022	2023	2024
Finistère	2	2	15	315	2139	6745	6728	7423	5903	5752	2981	8981	9229	
Côtes d'Armor	6	21	52	420	1600	3500	2423	3979	2750	3786	3490	8017	5893	
Morbihan	5	63	235	1147	2900	4933	3072	4159	2860	3570	2064	5031	6174	
Ille-et-Villaine	17	56	187	1122	3324	4850	3043	6200	2986	5042	2919	8061	8722	3851

*Nest numbers for the four departments in Brittany, France. At time of writing, only Ille-et-Villaine had released data for 2024, but weather adversely affected insect numbers generally in 2024, and YLH nest numbers are also expected to be low. The numbers in **blue** refer to years covered by the ITSAP 4-year study: Morbihan was a participant. Finistère, Côtes d'Armor and Morbihan all encourage trapping, while Ille-et-Villaine has not since 2018. Morbihan began co-ordinated spring trapping and nest destruction in 2015. Sources: FDGDON (2024); FGDON35, (2025).*

Environmental and Occupational Health & Safety) were not significant (in science, significance is a standard test to see whether the results could be down to chance). So we still have no clear proof that spring trapping works.

The GDS national plan

The Groupements de Défense Sanitaire (GDS), co-ordinated a 'national control plan' in France in 2022. This included 2 months of trapping for queens in the spring, using 'cone box' traps (like Jabeprode or Gard'Apis). In March 2023, the French Natural History Museum (MNHN) put out a memo saying that this trapping was not recommended outside of a scientific framework and should be limited to apiaries that had been heavily attacked in 2022. The MNHN pointed out that the spring trapping guidelines were based on the ITSAP/MNHN trial, but that the results of the trial should be taken with great caution because the statistical analysis was not complete, and the paper had not been published and therefore hadn't passed peer review. They point out that massive trapping campaigns could have a negative effect on insects and proper ecosystem functioning worse than the impact of the hornet itself.

Queen competition — natural population control?

In wasps, queen competition, known as 'usurpation' happens in the spring when an outsider queen challenges a queen in possession of an embryo nest. This is a well-known mechanism of population control, which the killing of hibernating queens or foundresses could disrupt. In New Zealand, when the widespread killing of hibernating wasp queens (*Vespa germanica* wasps were introduced into New Zealand in the 1950s) was encouraged by a bounty paid per head, the number of colonies later that year actually increased (Martin 2017). It is thought that this unexpected result was due to the removal of competition between queens. The only method that did decrease the density of wasp colonies in New Zealand was an almost complete destruction of nests: killing queens in the autumn or spring was completely ineffective.

M. E. Archer, who wrote the chapter on population dynamics in Edwards (1980), calculated the mortality rates of queens at various stages, and found a 98% loss of queens over the winter, presumably due to weather, parasitism (including fungi) and predation (although heavy rain and low temperatures did not seem to kill hibernating queens, and winter temperature did not influence wasp numbers the following year). A further 92% loss occurred in the spring due to bad weather, migration, predation, competition for nest sites and quality of queens (Edwards, 1980). In wasps, the usurpation of primary nests is very common and widespread, with at least 30% of colonies having undergone usurpation and an average of 12 queen changes in a nest (Archer in Edwards, 1980). There have been many scientific articles about this phenomenon in wasps, so is this also a mechanism of population control in YLHs?

Usurpation is often cited as a reason why the trapping of YLH queens in the spring to control the population may not be workable (Monceau & Thiéry 2017; they quote Haxaire & Villement 2010). The argument goes that if you don't try to trap the queens, then many of them will kill each other anyway. Conversely, if you trap queens, then the ones that don't get trapped will be more successful because they won't be challenged, and won't get injured. In a more recent paper (Thiéry & Monceau, 2024), they question whether queen competition can exist as a mechanism of self-population control if nests are being found at densities as high as 12 to 19 nests per square kilometre (Cotentin, France).

Martin (1992, 1990b) was able to observe usurpation by queens of the same species (intraspecific usurpation) in the hornets *Vespa affinis* and *Vespa simillima* in Japan, and noted that the vast majority of usurped nests failed (his own and others' observations). He thinks that usurpation is triggered by one of the queens having lost her nest. Usurpation is difficult to detect, and may be more common than we think; anecdotal evidence is starting to be gathered that it is occurring in the yellow-legged hornet in Europe. In China, where YLHs are raised for food, queens are kept strictly apart to avoid losses from usurpation (Van Itterbeeck et al. 2021).

It is clear that *something* naturally kills queens in the spring, or somehow prevents them from founding a successful colony that can reach maturity. Ille-et-Villaine gives the example of a small commune of 500 inhabitants where between 5 and 15 YLH nests are found each year. One spring, they caught nearly 1,000 queen hornets there in April and May; yet, again, no more than 15 nests were found later in the year. There's no way that this commune would ever have 1,000 nests (FGDON35, 2025).

Spring trapping conclusions

Where yellow-legged hornets are established, there is no strong evidence that spring trapping of queens reduces numbers of YLH nests and can control populations. Populations fluctuate naturally from year to year. When areas with and without trapping are compared, nest numbers are not noticeably different. Where spring trapping is practiced, so is nest destruction: can the effects of these two practices on nest numbers be disentangled? More data and observations are needed to find out what happens to the millions of overwintered queens that never found successful colonies. Is it down to usurpation, weather, disease, parasites or something else?

Large-scale spring trapping has gained traction because it feels better to do *something*, but we should be listening to the evidence. There is a place for trapping in an outbreak area (and not only in spring), for monitoring and possibly catching queens during eradication. In other areas, though, it is more useful to monitor during the summer and early autumn using a wick-pot so that nests can be tracked and destroyed.

Apiary trapping

Beekeepers need to be able to reduce the pressure from YLHs on their apiaries by using summer/autumn trapping. Shah and Shah (1991) in Kashmir, northern India, described using a fermented mixture of honey:water (1:1), to make a killing trap for severe predation. In 10 days, with 12 jars set out in an apiary, they caught 11,483 yellow-legged hornets (96 per jar per day), and the numbers of hornets hawking in front of hives went from 10–25 per hive to 0–3. Some UK beekeepers already use a similar technique for killing wasps around apiaries with a jam and water mix. There are some concerns about using honey and wax because of the potential for spreading bee diseases.

Rome et al. (2011), using a modified version of Shah and Shah's technique, found that apiary trapping close to hives could be moderately effective: where there was heavy predation (over 10 hornets in front of hives), 30–40% of the insects trapped were yellow-legged hornets (as opposed to non-target species), which is still pretty bad from a by-catch point of view. Flies and bees (presumably honey bees) were the main non-target insects caught, which is why it is important to place such traps as close to the hawking hornets as possible.

Koldo traps at the entrance to the hive are growing in popularity (see p 189).

Late autumn trapping for sexuals

Later autumn trapping is done to try to catch males and gynes (new queens) and is therefore done once the sexuals are emerging. We don't know exactly when this will be in the UK: in Jersey gynes started appearing in late September. Depending on how fast the winter comes, traps could be taken away perhaps in late November or early December.

Although later autumn trapping is preferred to spring trapping by ecologists (e.g. Monceau et al. 2013a), it generally appears to be the least popular form of trapping. The main argument for trapping at this time of year is that there are fewer insects on the wing, and therefore perhaps you could expect a lower by-catch. If traps based on mating pheromones become available, this will become the main trapping time.

Types of traps

Plastic bottle trap, dome trap and jar trap

Banned in the French national plan for dealing with yellow-legged hornets, plastic bottle traps essentially started as cheap drowning traps (plastic bottle traps cost practically nothing).The basic pattern involves a plastic bottle with the top third cut off and inverted to form a funnel trap. To reduce by-catch, modifications have been implemented which include: (1) making the entrance hole too small for native hornets but large enough for YLHs; (2) avoiding drowning insects by using attractant soaked into a sponge or paper, or placing a screen between the insect and the bait (as in the Andermatt trap); (3) allowing insects smaller than the YLH to escape (6-mm escape holes, ideally as high as possible; 7-mm escape holes should not let a queen escape); (4) having a way to take the trap apart to release insects; and (5) having a roof or cap to stop the bottle filling with rain and so drowning the occupants. The colour of the roof is important if the funnel is directly beneath it. The underside should be black so that trapped YLHs do not see the open funnel they came in through as an exit: they associate escape with light. Another version is a yellow plastic funnel with roof that replaces the lid on any 2-litre PET bottle (like a 2-litre pop bottle): this is also designed as a drowning trap (e.g. Tap Trap). Wasp domes where the insects enter through a wide funnel from beneath (e.g. Pest Trappa) are also designed as killing traps and are banned by the GDS in France when trapping yellow-legged hornets. Traps using 3D-printed cones screwed onto glass jars *may* also suffer from heating up in the sun, depending on amount of air flow (let me know!) and should be used with caution.

Any trap that is neglected will turn into a killing trap if insects cannot escape, whatever its design. Many insects cannot survive even a short period (a few hours) in a trap with plastic or glass walls like a plastic bottle trap, especially when there is liquid bait to drown in.

Experimental bottle traps. Photo by Chris Isaacs.

Even when the bait is kept separate with a grille, or absorbed into a sponge, heat, exhaustion, getting stuck in condensation, getting eaten by a predator in the trap, and getting covered in sticky material are all factors leading to insect deaths.

It is also important to note that just because a trap has some exit holes, it doesn't mean that non-target insects are able to escape through them: they may not find the holes, or recognise them as escapes, or be able to pull themselves through.

The main design problems with plastic bottle traps are that insects naturally climb up and get stuck between the funnel and the walls of the bottle, and the bottle heats up easily in the sun, making it too hot for insects and forming condensation when it cools.

Lastly, one study (Renoux et al. no date given, but working for the GDS) compared three types of trap: plastic bottle with screw-on funnel (Tap Trap), Véto-pharma's Vespacatch and the Jabeprode with a solid wooden box. They found that the VespaCatch killed one YLH for 118 by-catch insects; the bottle trap was the same, and the Jabeprode killed 1 YLH for 22 by-catch insects.

The Véto-Pharma trap

The Véto-Pharma trap is essentially a manufactured version of the home-made plastic bottle trap. Originally designed as a drowning kill trap, it is used in Jersey as a monitoring trap with a mini wick-pot inside, or as a bait station with the entrance cones removed.

This is the trap used in Guernsey for their annual spring queen trapping, with modifications made over the years, from the barest amount of bait soaked into a sponge

Roof of Véto-Pharma trap peeled back to show lid with entrance funnels: the lid can be removed to make a bait station.

in the bottom, to two sets of 6-mm holes drilled around the top of the side walls, and just above the sponge. The real modification that has led to ever-smaller amounts of by-catch death is the army of committed volunteers who release by-catch every day.

Mesh and cone traps

Raoul's cage and the Jabeprode (stands for Denis le **Ja**ffré's **be**e **pro**tection **de**vice) are box traps, in which perforated entrance cones lead hornets inside looking for bait that is kept separate under a mesh floor. The boxes are solid, with small areas of mesh wall attached to the cones. Replacing some or all of the box walls with mesh that allows other insects to come and go (unless they get eaten by the hornets) should significantly reduce by-catch, but Denis Jaffré reckons that hornets are then less likely to find their way in through the cones as they get distracted by the

The Jabeprode. Top, inside showing cones and trapped hornets. Above, a homemade version using the Jabeprode cones that can be bought separately. Top photo: www.jabeprode.fr. Bottom photo: Barry Duke.

The Gard'Apis trap, outside and inside. The trap comes with two different sized entrance nozzles, a larger one for catching queens and a smaller one for catching workers. It is supplied with little pots that can be turned into wick-pots, which avoids spills. www.gardapis.co.uk

*The Andermatt traps. On the **left** is the YLH live trap, sometimes called the 'blue and yellow trap'. Hornets enter through the lid, and there are escape holes large enough for wasps to get out, also in the lid. On the **right** is the orange YLH trap. This trap comes with a Tuppaware-like lid, too. Hornets enter through the orange cones and other insects can escape through the sides of the cones, or through the grille at the top. www.andermattgarden.co.uk*

The Belgian/Dutch trap. A wick-pot is used to attract hornets. Cone entrances are 3D printed. Photo Dominique Soete.

bait smells coming through the additional mesh (Denis Jaffré, pers. comm.). One might think that a mesh ceiling would be good for encouraging by-catch to escape, but a downside is that rain can flood the bait and box.

The Belgian/Dutch trap is a similar box-and-cone trap to the Jabeprode, but home-made, and uses a wick-pot instead of a bait tray under the trap, separated by a grille. This simpliflies the construction. Both Jabeprode and Andermatt sell cone entrances separately, that can be used to construct your own trap, but builders of the Belgian/Dutch trap tend to use 3D-printed entrance cones. If you have access to a 3D printer, the files can be downloaded from the Internet. Search for '3D piege frelon'.

The most recent traps to come onto the market are the Gard'Apis and two traps sold by Andermatt. The Gard'Apis uses French plastic queen excluder material formed into a cylinder, with large entrance cones on either end. The queen excluder allows by-catch to escape all the way around the cylinder. They have gone to a lot of trouble to get the entrances as selective as possible by having two sets of nozzles that close the cones down to 8 mm for catching queens in the spring, or 7 mm to catch workers in the apiary later in the year.

The Andermatt blue and yellow trap is based on the plastic bottle trap, with a mesh floor separating the bait from the main chamber, and an entrance cone at the top. The orange trap has entrance cones in the side walls and a mesh ceiling, which may be easier for by-catch to get out, but will let rain in.

In the spring of 2024 (mid April to mid May), three different types of 'selective' trap (modified Véto-Pharma with a wick for bait release and six 7-mm escape holes; the Gard'Apis; and the Jabeprode) were tested along with a Véto-Pharma kill trap by a team from Exeter University in Cornwall. With traps checked every week, they found that these 'selective' traps still caught a wide range of by-catch at an average of around 5-8 insects per trap per week, and the kill trap caught more than 150 by-catch insects per trap per week. This, in a spring that was noticeably poor for insect abundance. So although by-catch is reduced versus kill traps, it isn't zero for selective traps.

Sticky trap

Finally, a trap I barely want to mention, the sticky trap. Basically, it is a board coated with an extremely sticky glue. A special bait, or a live yellow-legged hornet is used to attract hornets onto the gluey surface. Unfortunately, anything landing or walking on the surface cannot be released and so is doomed. A beekeeper in France who tried it out found that, alongside hornets, mice, lizards, European hornets and moths were caught, so stopped using it (Kevin Baughen, pers. comm.). If lizards get caught on it, small birds might also get stuck. I hope that this is never available in the UK.

Type of attractant

When we look at the natural history of the yellow-legged hornet, we see that, theoretically, different foods should be more attractive at different times of the year (see graph p 58). When the foundress wakes up from hibernation, she should seek sweet carbohydrate types of food such as sap and nectar, and therefore sweet attractants should work best. Once her first batch of eggs hatch into larvae, she will need to find protein for them (insects, carrion), and will receive larval exudate from the larvae once they are old enough. She is also likely to take advantage of other sources of sweet liquid carbohydrates when she is out and about.

Once the first workers emerge from pupation, they will start foraging to get protein to feed the larvae in the nest. Some authors suggest swapping from a sweet attractant to a protein-based attractant (raw fish) from the beginning of August to mid October (France; Monceau & Thiéry, 2016). But workers still find it hard to resist a sweet attractant (after all, they still need sweet carbohydrates for energy). In Jersey I saw yellow-legged hornets feeding on Trappit in a dish placed between two hives, even though active honey bee hawking was going on in the apiary at the same time (ideal refuelling to carry on the hunt?). In the autumn, when the sexuals leave the nest, they first feed as much as possible on larval regurgitations, which are a perfect mix of carbohydrates and proteins for the adult insect. Autumn traps should concentrate on sweet bait once again. Protein (e.g. skinless chicken) in traps goes off and gets smelly. At some point it will no longer be attractive to hornets and will attract blowflies (e.g. bluebottles) instead. Fish and shrimps have been tried out with some success in France, Jersey and the UK.

Lim et al. (2019) found that a bacillus strain, *Bacillus* sp. BV-1, grown on sugar medium, attracted yellow-legged hornets. The volatile compounds that could be attracting the hornets were: 2-methyl-1-propanol, 3-methyl-1-butanol (probably the main attractant), 3-methylbutanoic acid, ethyl hexanoate, 2-phenylethanol, ethyl octanoate and ethyl decanoate.

Two yellow-legged hornet workers feed on fondant. Photo by Angus Deuchar.

Pheromone trap

Work is ongoing in designing an attractant that is based on yellow-legged hornet pheromones. If this is successful, then a truly selective trap, which will only attract YLHs, can be manufactured: the holy grail of trapping. Cheng et al. (2022) found that a blend of the sex pheromones 4-oxo-octanic acid (4-OOA), 4-oxo-decanoic acid (4-ODA), and 5-oxo-decanoic acid (5-ODA) in the ratio of 4-OOA:4-ODA:5-ODA 2:4:1 were attractive to both *Vespa velutina nigrithorax* and *V. v. auraria* males. However, visual cues, such as a dummy female are also important, and it wasn't found to be as effective in France as in China, possibly due to the control measures such as nest destruction that have taken place in France. There are still several puzzles to work out before this pheromone would be ready to commercialize, but it is promising.

Cheng et al. (2017) found that yellow-legged hornet venom acts as an alarm pheromone, which will attract other hornets. Beekeepers in France regularly use live trapped YLHs to attract others.

If I kill a foundress queen in the spring, that's equivalent to one nest removed, and therefore I will stop huge numbers of insects from being killed later in the year, right?

Not necessarily. If every queen emerging in the spring went on to make a nest, we would be inundated by them; but most queens that survive the winter don't go on to found nests (see the example of Ille-et-Villaine, p 165). During **eradication**, it is worth trapping for queens within a 5-km radius of a nest that was taken down late enough for sexuals to have emerged, or a suspected, but unfound nest. It is also worth trapping in high-risk locations like ports. Where YLHs have become **fully established**, trapping of foundresses in the spring ('spring queen trapping') does not seem to affect the number of nests found later in the year.

I think I've found a yellow-legged hornet on my bait station/in my monitoring trap — what should I do?

If you can, take some photos of it in the trap or feeding at the bait station straight away (they are not aggressive when feeding and a long way from the nest). If it's in a trap, use an identification guide (this book!) to make sure it isn't a native European hornet or one of the other insects it may be confused with. If you are sure it is a yellow-legged hornet, then put the whole trap into a plastic bag and put it in the freezer overnight. Once dead, you can tip it onto a surface and take better photos. Keep the dead hornet in a small container in the freezer: it may be useful for the National Bee Unit. Use the Asian Hornet Watch app on your smartphone to report it.

If you would like some help with identification, contact your local Asian Hornet Action Team, which is part of your local Beekeepers' Association. To find out where your nearest AHAT co-ordinator is, there is a map on the British Beekeepers' Association website (www. bbka.org). On the main page, scroll down to 'Asian Hornet' and click, then choose 'Asian Hornet Teams'. A map will come up and you can put your post code in or just find where you are on the map. If you then click on surrounding blue pins they will show the names and contact numbers of nearby AHAT co-ordinators.

Trapping conclusions

You can can use selective traps to monitor for YLHs if you are in a red zone. But please only trap if you can commit to releasing by-catch daily, if not twice daily. Otherwise, monitor with bait stations.

Remember that the by-catch insects caught in spring traps are often different to the insects killed by YLHs later in the year, and many are too small to have been targeted by hornets (Rojas-Nossa 2018). Also, traps work all day and all night (hence finding night-flying moths in traps), whereas hornets only forage during the day, so again by-catch insects killed are in **addition** to those killed by hornets. Flies, wasps, moths and others serve specialist functions in our ecosystem, and their loss in traps is not justified in areas where there is a very low chance of seeing a yellow-legged hornet.

A honey bee guard confronts a yellow-legged hornet. Photo by Sylvie Richart-Cervera / INRAE SAVE.

Other Approaches for Control

Poison

Vespids (the social wasps, including hornets) have been successful in establishing all over the world: sometimes introduced by humans, on purpose or accidentally, sometimes spreading naturally across continents. The social wasps (those that live in colonies, with only some capable of reproduction, like yellow-legged hornets) seem to be particularly good at invading new areas, and their impact can be large and wide-ranging (Beggs et al. 2011).

So far, toxic baiting has been the most successful control strategy against invasive vespids. Workers take the bait back to the nest and feed it to the larvae. Insecticide-laced bait has been successfully used to reduce populations of wasps all over the world (Patagonia in Argentina, Tasmania in Australia, Hawai'i in the USA, and in New Zealand). The insecticide used is fipronil, and it is mixed with minced meat at a concentration of 0.1%. Using the commercial version of this bait (Vespex®), wasp traffic was reduced by 97% after around a month (Edwards et al. 2017). In their paper that looked at the transition from field trials to larger areas, they were able to bait plots from 217 ha to 2477 ha in New Zealand against *Vespula* wasps. Nests were not removed after poisoning — indeed on these large scales, they didn't even know where they were. However, the New Zealand Department of Conservation managed the sites where the bait was tested and the New Zealand Environmental Protection Authority approved the use of Vespex® in New Zealand.

The problem in Europe is that there are plenty of scavengers (insects which will eat meat or meat juice) that are non-target insects, including wasps and the European hornet.

In France, people have resorted to mixing up a home-made poison, using fipronil, egg, flour and water, and painting it onto captured workers. The workers are then released and fly back to the nest, poisoning the adults and larvae in the nest. In Spain, a similar method, which they call the 'Trojan method', involves spraying a worker with insecticide, removing its back legs to prevent it from grooming, and letting it go back to the nest. A Spanish company called LOYDERN manufactures little capsules that can be attached fairly easily to a foraging worker. The capsule is filled with wasp-killing powder, capped with e.g. a thin piece of potato, and the worker carries it back to the nest. As the potato dries out, it shrinks, releasing the poison. Presumably this would need to be done shortly before dusk so that the hornet then stayed in the nest for as long as possible, rather than continuing to forage. The problem with all these methods is that the nest is not found or removed, and once it is undefended, birds, small mammals and other insects will move in to feed on dying adults, pupae and larvae, potentially causing a cascade of poisoning in the local ecosystem.

Barandika et al. (2023) conducted a three-season trial into protein bait laced with fipronil (0.01%: ten times lower than the amount used to control wasps in New Zealand) in apiaries in Spain that were under attack from yellow-legged hornets. In the lab, the poisoned bait killed all larvae in 48 hours (apart from those larvae starting to spin pupal cocoons — they were not interested in food). Bees were shut in during the bait-feeding sessions in the apiary. The trials included 222 beekeepers, and the data from the 90 beekeepers who completed the questionnaire showed that those who were suffering from the highest pressure of hornets (those with 10–30 hornets and those with more than 30 hornets) had a significant decrease in hornets, lasting at least 2 weeks. Those beekeepers with fewer than 10 hornets in the apiary found a slight decrease in hornets, but this quickly recovered.

The authors note that the degradation residues of fipronil are extremely toxic to bees and other non-target species. If this method were to be used, it should be carefully restricted, with maximum consumption of the bait in the shortest possible time.

Fipronil can take years to break down, especially if it is in the dark. It has a half-life (time to 50% of poison remaining) of 4 months to 1 year. It is highly persistent in soils, tends to accumulate in soils and sediments, and has a high run-off and leaching potential to surface and groundwater (Pisa et al. 2014). Fipronil is extremely toxic to bees and other organisms. It is used as a soil treatment and seed dressing, but was banned from use on maize and sunflowers from the end of 2013 in Europe after a mass die-off of bees in France was blamed on fipronil-dressed seeds.

In Japan (Kishi & Goka 2017) where, like the UK, they are expecting yellow-legged hornets to establish on the mainland, they have been looking at chemicals that could be used against them. They investigated 22 possible compounds: of the six most effective chemicals, three are now prohibited in most countries, and one was highly toxic to mammals. This left them with fipronil and diazinon:

the former can harm important pollinators and other non-target invertebrates, while the latter is an organophosphate insecticide, suspected of being carcinogenic. Looking for alternatives, they are investigating insect growth regulators (IGRs), which strongly affect the development of eggs, larvae and pupae, but do not kill adults immediately. They have had some success with etoxazole, which suppressed the pupation of Japanese native hornets in the laboratory and caused colonies to collapse in the field. They are continuing with field trials, but again, no-one knows the effects of these chemicals on non-target organisms.

Yellow-legged hornet seeking nectar from a Camellia sasanqua *Navajo, on Jersey. Photo by Peter Kennedy*

Biological control

In their native Eastern China, yellow-legged hornets are medium-sized hornets, and their greatest predators are bigger hornets (for example the Asian or northern **giant** hornet, *Vespa mandarinia*) and humans, who kill them for food, protection of apiaries and protection of communities. Even though there are several mammals and birds that will opportunistically eat yellow-legged hornets, (see p 55), the only predator likely to have a real impact in Europe is the European honey buzzard, (*Pernis apivorus*). In northwestern Spain, Rebollo et al. (2023) found that the 700 breeding pairs of honey buzzards could have destroyed as many nests (24,500) as destroyed by the local authority in 2018: an extraordinary impact.

Although their predators are few, we all have smaller things out to get us, and so attention has turned to European parasites that could help bring YLHs under control. When organisms are transplanted into a completely new environment, they may be freed from the parasites from their native range, but be susceptible to parasites in their new territory. They may be vulnerable to diseases and parasites carried by insects they come into contact with, especially honey bees, European hornets, wasps and other prey.

However, it may take time for native parasites to adapt to invasive species, and in the meantime, the pressure taken off alien species by leaving their parasites behind can mean that they can put more resources into growth and breeding: this may be one of the factors involved in the success of YLHs in Europe (Darrouzet et al. 2014).

Conops vesicularis

In the summer of 2013, twelve primary YLH nests were observed on buildings in and around Tours, central France (Darrouzet et al. 2014). Of these, only three developed normally, showing what a hazardous time foundation of colonies is for foundresses. Of the colonies that died out, seven had queens which could not be found, and two had dead queens in them. These queens were dissected and both were found to have been parasitised by an extremely specialist type of fly, *Conops vesicularis*, native to Europe.

Conops vesicularis. *Photo by By Tristram Brelstaff (Own work, CC BY-SA 3.0).*

They fly from May to September and wait for a host on a flower before grabbing it and laying an egg into its abdomen. The larva develops inside the host (a process that eventually kills the host) and exits the host when it becomes an adult. The fly is unlikely to affect YLH populations by attacking foundresses because foundresses themselves undergo massive reductions in numbers in the spring. Also, *Conops vesicularis* rarely attack social wasps, but prefer bumblebees. Therefore, using these flies as biological control agents could be too risky to native insects to contemplate.

Other insect parasites

There are a few insects that parasitise other hornets that may be worth pursuing. *Xenos moutoni* (family Stylopidae) parasitises *Vespa analis* in Japan (Makino et al. 2011); when parasitised, the hornet is unable to reproduce. Kanzaki et al. (2023) collected several species of hornets (not *Vespa velutina*) from baited traps, and by hand, and found that parasitism by *Xenos* species varied between 0 and 17%. Interestingly, the parasites were more common in hornets from traps, as it seems that *Xenos* affects its host's behaviour, making them more likely to visit and stay at feeding sites than forage or reproduce.

Sphecophaga vesparum (family Ichneumonidae) has been found to parasitise *Vespa orientalis* in Israel (Turchi & Derijard 2018). The females enter the nest and lay their eggs on top of the host's larvae or pupae. The parasitic larva then attacks and feeds on the host's larva and then continues its development in a cocoon in the nest.

Volucella inanis is a hover-fly (family Syrphidae), and part of an amazing genus of hover-flies which mimic and parasitise different social hymenopterans. The hover-fly larvae enter the cells of their host and eat the host's larvae.

Recently, a moth larva that destroys bee and wasp comb, *Aphomia sociella*, was found on yellow-legged hornet comb in Jersey, by Chris Isaacs. Again, using such an insect as a biological control could impact wild bumblebee or wasp populations.

Bee moth larva, Aphomia sociella, *on yellow-legged hornet comb. Photo by Chris Isaacs.*

Nematodes

In a 2015 paper, Villemant et al. discussed the possibility of mermithid nematodes being used as a biological control for yellow-legged hornets. The species is *Pheromermis vesparum* — one which specialises in social wasps in Europe. The only thing is that, in 10 years, only three nematodes have been found in YLHs, despite scientists handling thousands of hornets (33,000 hornets from 77 nests). This is probably due to its complex life cycle. Adult nematodes lay their eggs in water, and these are later consumed by larvae of aquatic insects, such as caddis flies or stone flies. The eggs hatch in the digestive tracts of their hosts, then move into tissues outside the digestive tract. Incredibly, they survive metamorphosis of the host into its adult form, inside a tough covering, and this encysted larva is transmitted to a hornet larva when the latter is fed the flesh of the adult caddis fly. The nematode, in classic parasite style, eats the non-vital tissue of the hornet larva, somehow survives another complete metamorphosis of this second host, and is then found in the adult hornet as a nematode worm almost 10 cm long, close to maturity. When the host hornet approaches water, it breaks out, killing its host. When the diet of yellow-legged hornets was analysed, caddis flies represented only 0.2% of insect prey, which is perhaps why the nematodes are found so rarely.

Another two species of nematodes, *Sphaerularia vespae* and *Sphaerularia bombi* have been found to parasitise gynes of *Vespa simillima*, resulting in the queens being unable to reproduce. In Japan, more than 60% of *Vespa simillima* gynes collected were infected with these nematodes (Kanzaki et al. 2007). Both of these nematodes are present in France, so there is potential for them to be used as biological controls, but they could also infect bumblebees.

Finally, *Steinernema feltiae* is a nematode which is already used as a biocontrol in France against caterpillars and ants and has been tested against wasps: there is some potential here, too (Turchi & Derijard 2018).

Entomopathogenic fungi

Entomopathogenic fungi are those that specialise in attacking insects: the fungal spores land on the insect and, if conditions are right (especially temperature and humidity), they can germinate and their hyphae (thread-like 'roots': they search for nutrients, build fruiting bodies etc.) can travel right through the insect, killing it in the process.

Poidatz et al. (2018b) have looked at using these fungi for yellow-legged hornet control. They concluded that perhaps the best way to use the fungus would be to spray a suspension of spores into a nest, so replacing the broad-spectrum insecticides that are used for nest destruction at the moment. Although whether such a method could infect non-target insects needs more investigation, the authors think that this risk is low. In a more recent paper (Poidatz et al. 2019), they also discuss the possibility of infecting yellow-legged hornets as they access bait; the infected insects then return to the nest and infect the colony.

Carnivorous plants

Christian Besson, gardener and botanist at the Nantes Botanical Gardens, France, first noticed many yellow-legged hornets were getting trapped by *Sarracenia* — a type of North American pitcher plant. Meurgey & Perrocheau (2015), from the Natural History Museum and Botanical Gardens, respectively, in Nantes, then cut open 203 pitchers to find out how many were being caught. Another study (Wycke et al. 2018)

Sarracenia. *Photo by Noah Elhardt (CC BY-SA 3.0).*

found that yellow-legged hornets accounted for only 4.3% and 0.7% of all insects caught in pitchers in 2015 and 2016, respectively, although there were several YLH nests nearby. The vast majority of insects caught were flies. So, it appears that they are not selective enough, and the plants themselves are tricky to cultivate.

Chickens

Job Opportunity — for self-confident chickens with good eye-beak co-ordination, there are several permanent positions available in an apiary near you! Some beekeepers in France have noticed that the free-range chickens which forage around their hives actually pluck yellow-legged hornets out of the air when they are hawking in front of hives. I don't know how they avoid the venom from the hornets, or how many they can deal with in a day. But since the combination of orchard, apiary and chickens is quite common, it's worth a try.

By the way, don't expect chickens to deal with a nest in an enclosed space — if the hornets build a nest in the chicken coop, a chicken caught in there with them can easily be stung to death (Kevin Baughen, pers. comm.).

Other pathogens

As mentioned previously, Gabín-García et al. (2021) found that yellow-legged hornets harbour most common hymenopteran pathogens, including Trypanosomatidae (flagellate protozoa), Nosematidae (microsporidians) and viruses. Interestingly, most of the hornet samples were parasite-free, which is strange if they are visiting the same flowers as other species that carry parasites, and also predate parasite-carrying insects. The authors wonder whether this could be due to subtle differences in feeding habits, such as: (1) perhaps not visiting the same flowers, (2) not actively harvesting pollen, which could be contaminated by other insects; and (3) a common practice of discarding the abdomen of prey may reduce their exposure to gut pathogens. Suffice to say that sampled hornets do not seem to be particularly susceptible to European pathogens, and use of pathogens against them could have bad impacts on closely associated insects.

Dalmon et al. (2019) found that, among 19 detected viruses found in the yellow-legged hornet, Deformed wing virus B (DWV-B) was predominant in all the samples, particularly in the muscles of a hornet with deformed wings. DWV, ABPV (acute bee paralysis virus) and BQCV (black queen cell virus) have been found in the brain and muscles of yellow-legged hornets, showing that they are actually infected, and not just carrying the virus. As well as seven species of virus already known in the honey bee, they also found 11 new species. Two new hornet viruses (*Vespa velutina* associated acypi-like virus and *Vespa velutina* associated triato-like virus) also seem able to multiply in the hornet, and could be candidates for biocontrol. However, much research would need to be done in order to find out whether this would work and be safe for other insects.

Mites

Very few mites are found in vespids (Turchi & Derijard 2018), although a handful of yellow-legged hornets have been found carrying *Varroa destructor*, the mite that causes a lot of problems for beekeepers. Recently, researchers in New Zealand have isolated *Pnemolaelaps niutirani* in their search for biological controls against wasps: perhaps a mite could be used against *Vespa velutina*.

Integrated pest management

Robinet et al. (2016) suggest using combined strategies of mechanical removal and destruction of nests alongside biological control and pheromone traps to slow down the spread of the yellow-legged hornet and reduce its population density at a large scale. They also point to the inherent riskiness of biological control, which can go horribly wrong due to unforeseen consequences.

Genetics

As we have seen, the ability of a single mated queen to start a colony, and therefore an invasion, has been highly beneficial for some wasps and hornets. But such an event produces an extreme genetic 'bottleneck': a large amount of genetic diversity is lost.

Arca et al. (2015) looked at the French and South Korean populations of *Vespa velutina* and found that they had much less genetic diversity than the native populations in China, indicating a severe genetic bottleneck caused by single founder events for these new invasions. Despite such a narrowing of genetic diversity, yellow-legged hornets have conducted a spectacularly fast invasion of Europe, due to a favourable climate, easy food supplies in the form of the defenceless local honey bees (*Apis mellifera*), possibly a lack of parasites they would normally be exposed to and, importantly, minimal competition from native hornets.

So far, there isn't much evidence for detrimental effects of the bottleneck. One possible indicator is the production of diploid males: eggs that should have developed into females develop into males instead, due to a lack of diversity of genes for sex allocation. When a lot of diploid males are produced very early in the season, the colony may fail due to lack of help in feeding larvae and nest building.

Selection of honey bee strains

In Europe, the docility of honey bees is a top trait that has been selected for, alongside colony health and productivity. Maybe we will have to rethink this, at least in terms of aggression towards yellow-legged hornets. As we have seen, the honey bee prevalent in *Vespa velutina*'s home territory, *Apis cerana,* displays defensive behaviour which helps the colony survive, including heat-balling, shimmering and different evasive flight patterns around the hive. Perhaps in the future we will have to consider selection of colonies which are able to cope with the pressure of yellow-legged hornet predation (Monceau et al. 2014a).

DNA technology

A possible high-tech solution to *Vespa velutina* control would be to interfere with its gene expression, for example by stopping it from producing some vital component used for its development. First, suitable genes that could be interfered with would need to be found and understood, and then a delivery system would need to be developed. Because the larvae are fed meat, the production of a genetically modified fish, for example, which produced specific RNA interference molecules that would interfere with larval development, could be developed as a biologically toxic bait. Such an approach could be extremely specific (Turchi & Derijard 2018). In a study in 2020, Wang et al. investigated functional target genes and molecular markers in the brain of *Vespa velutina* and found plenty of genes that could be focused on for future control projects.

Meiborg et al. (2022) investigated the potential of using gene drive technology to control wasp populations, including the yellow-legged hornet and the European paper wasp. This involves introducing a genetic element that spreads rapidly through the population, but which causes problems once it is prevalent. Such technology could have far-reaching unintended consequences. For example, if *Vespa velutina* carrying a modified infertility gene were to get into the native population back in China, it could have massive knock-on effects on the ecology there, where several different species of hornets at present co-exist. Much recent research has concentrated on making this type of technology safe and controllable, once the implications were understood.

The authors modelled the life cycle of a typical haplodiploid social wasp, and what would happen if a gene drive was introduced. They found that such a genetic biocontrol method didn't really work with these social hymenopterans, because of the way females and males are determined. They conclude that unless much more efficient gene drives are developed, conventional approaches such as nest destruction remain the best control options.

Beekeeping with Yellow-Legged Hornets

Introduction

As beekeepers, how will we cope if yellow-legged hornets become established in the UK? The first thing to say is that beekeepers still carry on elsewhere, and have had to adapt by using a combination of simple interventions along with different management practices. The most important tool is understanding how the yellow-legged hornet works: understanding the life cycle of this hornet will show you when in the year it will be a threat to your bees, and why, and also what you can do to out-wit it. Every piece of knowledge about this hornet will give you vital clues as to how to work against it.

In this Chapter, I will discuss physical interventions and management practices that may have to become normal if we are to live with this hornet. First of all, however, I want to talk about stress.

Stress

Stress is instrumental in the response of honey bee colonies to yellow-legged hornet attack. This has become clearer over the years, and reducing stress on colonies is paramount when dealing with the yellow-legged hornet. The stress that bees are under can be lowered in two ways: by reducing the impact of hornets on your colonies (using various interventions), and by increasing the resilience of your bees through good apiary management. Partial or complete foraging paralysis is disastrous for bees because it leads to a downward spiral of colony collapse (see p 67).

The best way to reduce the effects of hornets on your bees is to get rid of them completely, or at least the vast majority, by finding nests (using simple tracking methods outlined in this book) that are within 700 m, or even a kilometre radius around your apiary, and getting them destroyed. If that is not possible, then you will have to deal with the hornets as they arrive at the apiary.

Physical interventions

Move the hives

First, if you are suffering from hornet attacks, do you have the option of moving your bees somewhere else that is free of hornets? This is the simplest thing you can do, and in the years between hornets becoming established and their numbers building up, there may be ample opportunity to do that.

Reduce numbers of hornets

We have mentioned **trapping** of hornet foragers, especially with mesh box and cone traps, to reduce the numbers of hornets hawking in the apiary (p 165). Traps in amongst hives are less likely to catch non-target insects. If there are several nests nearby, numbers of foraging hornets can be staggering. A medium-sized hornet nest in mid September may have around 2,000 adult workers. Now you can see why large-capacity traps, such as the Jabeprode, are the choice of those under high hornet pressure in the apiary. However, one ITSAP study (Decante 2015) found that apiary traps were not able to save colonies from the stress of predating hornets, either under high or low hornet pressure.

Jabeprode hornet trap set up in apiary. www.jabeprode.fr

A different trap is becoming popular on the Continent, the **Koldo trap**, named after the inventor, Koldo Belasko, who keeps bees in northern Spain. It is made from plastic (4 mm × 4 mm) or galvanised mesh, or fabric netting, and is attached over the entrance to the hive. There is an opening that bees can use to leave and return to the hive, and this entrance also attracts YLH. But once inside the device, hornets, feeling enclosed, tend to fly upwards and end up entering a 'chimney' that protrudes from the top. The chimney has a cone entrance that effectively stops hornets from being able to escape back towards the hive. These traps can collect vast numbers of hornets. Some people attach one trap to each hive, but Koldo has also experimented with a 5-m tunnel made from 6 mm × 6 mm galvanised mesh that runs along in front of several hives, with entrances around every 1.25 m. The tunnel leads into a Koldo trap. Bees are able to get through the galvanised mesh, but hornets fly along the tunnel, eventually ending up in the chimney of the trap.

A Koldo trap used in Spain. The box of the trap is made from galvanised mesh, and the 'chimney' from plastic mesh. The chimney is removable to kill and empty the hornets within. Photo by Ester Ordoñez.

Quite a few French beekeepers use swatting with a **plastic racquet**, followed by a judicious stamp to destroy hawking hornets. However, if there are a few big nests in the area, you will hardly make an impact unless you are able to kill hundreds in a day, for days on end. The electric racquets designed to despatch mosquitos also look satisfying, but the hornet is only stunned and will again need to be stamped on to kill it. The downside (apart from arm-ache) is that constantly waving a racquet in front of the hive is likely to annoy your own bees so much that they may come after you!

SANVE wet electric harp. Vast numbers of yellow-legged hornets have drowned in the water, which needs to be scented (e.g. lemon washing-up liquid) to keep bees away. The solar panel charges a battery that powers the electric harp.

The **electric harp** ('harpe electrique' or 'arpa electrica') is another way to kill yellow-legged hornets in the apiary. It is a frame that contains a lot of fine vertical wires that are electrified. It is placed to intercept the flight paths of hornets that tend to circle hives looking for prey or a good spot to hover. They can be placed in different ways (see illustration). It is powered by a car battery that can be swapped out, or charged by a solar panel with a specific controller. When a hornet flies into the harp, it is stunned by an electric shock, and drops down, either into water, where it drowns, or into a dry trap at the bottom. One YouTube video shows a catch after a week of 1,170 yellow-legged hornets, no European hornets and 14 bees. In France

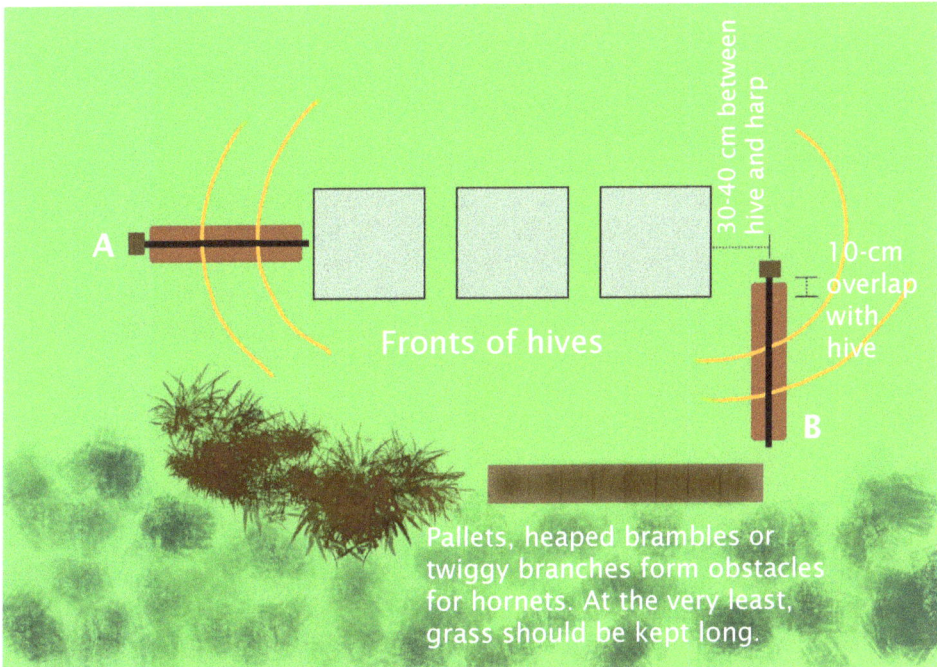

*Placing of electric harps according to manufacturer SANVE. Two different positions for **summer** apiary defence are shown (usually beekeepers prefer one position over the other). **A**. At side of hives. **B**. Sticking out in front of hives. The obstacles force the hornets through the harp. No attractants are used, and no muzzles or other impediments for the bees. The yellow lines show flight paths of hornets.*
*In **spring** and **autumn**, the harps are baited with attractant (sugary/fermenting in spring and beeswax in autumn), and the harps are placed well away from the hives to catch queens in spring and sexuals in the autumn. sanve.weebly.com*

the harp has won three awards. Electric harps can be bought (see Resources) or made at home: if you are a member of the BBKA, Andrew Durham's detailed making instructions are available from the BBKA website, or you can search online using 'Asian hornet electric harp' or 'harpe electrique pour frelon'. But don't attempt building your own unless you are confident with electrics, as it uses dangerously high voltages.

Rojas-Nossa et al. (2022) found that electric harps reduced predation pressure and therefore lowered foraging paralysis. Colonies that were protected by electric harps were able to forage, brought in more pollen and had higher brood production and worker body weights. This led to better winter survival, especially at apiaries under intermediate to high hornet pressure. The authors thought that they should be used alongside other management strategies, such as feeding during periods of high predation, as well as finding and destroying nearby nests, since the harps could not completely stop hornet predation. The harps were fairly selective in that more than 90% of the insects captured were yellow-legged hornets, followed by other hymenopterans, such as bees, then large flies.

Thiéry et al. (2023) also looked at electric harps and came to similar conclusions as Rojas-Nossa et al. They used the harps in similar positions to the previous study, setting them up in late August and catching 9,287 yellow-legged hornets between four harps, over 47 days. The hornets represented 80.61% of total captured insects, with honey bees coming in at 16.23%, and other insects less than 3.5%. Honey bee losses were put down to them seeking water and drowning in the water containers that were part of the electric harp traps. The authors tried adding 20% vinegar or 10% salt to the water to discourage them, but it only marginally reduced the numbers of honey bees caught. Intriguingly, the end harps caught fewer hornets than the harps between hives, but this was not statistically significant. It would be interesting to see whether this is a real effect in future research. In an article by Pérez-Granados et al. (2023), similar results were found, with dry harps (where hornets collect in a container and smaller insects can escape), being more selective than wet harps. Electric harps can need a lot of attention, especially in terms of repositioning (the hornets learn where they are) and reactivation when they stop working.

SANVE wet electric harp. Yellow-legged hornets drown in the water.
sanve.weebly.com

Disrupt hawking

Hawking (the hornet behaviour of hovering in front of hives and attacking flying honey bees as they return to the hive) can be made difficult by allowing grass to grow long in front of the hive entrance, but as well as providing interference for hovering hornets, if such measures force bees to land away from the hive and carry on by foot, they can also become easy targets. On the other hand, it provides somewhere for bees to hide too, and overall does seem to help. If you are going to use electric harps, then the grass needs to be short beneath the harps and in front of the hives, but should be left long beyond that, to force the hornets into the electric harp zone (see illustration).

If hives are up on stands, the areas under the hives should be screened off to stop yellow-legged hornets from using these areas to wait for returning foragers. In addition, hawking can be disrupted by piles of thin branches ('brashings') or brambles in front of the hive. Beyond this simple intervention, some kind of mesh, net or cage can be used to disrupt hawking at the hive front.

The holes in the mesh need to be big enough for bees to fly straight through, but small enough that hornets will hesitate because of the size of the gap and the fact that they would be entering the colony's territory. A net or cage with smaller holes that the hornet can't get

A home-made 'muzzle'. This one has solid sides and floor. Photo by Bob Tompkins.

through means that the bees have to land on the mesh and then crawl through, making them an easy target. However, if the area of netting is large enough, this may not be a problem. Research is needed on this question.

The most popular mesh arrangement is the 'museliere' or muzzle, a wire mesh or plastic guard that covers the landing board of the hive. There are many variations of this device because they are often home-made. The typical mesh used is around 25 mm (1 inch): apparently chicken wire works just as well.

The 'STOP-IT' muzzle, a French product that is available in the UK. The one on the **left** has a single barrier, while the one on the **right**, the STOP-IT Max, has an extra barrier inside with holes too small for the yellow-legged hornet to get through. www.stop-it.fr

Requier et al. (2020) looked at the effect of using muzzles on hives. Unfortunately they used muzzles with a very small mesh (6 mm square mesh), which forces bees to crawl through. Even so, they found that the muzzle led to a reduction of up to 41% in foraging paralysis. Survival of hornet-stressed colonies increased if they had muzzles. So although the muzzle did not stop predation, it drastically reduced foraging paralysis. The authors did suggest larger mesh, and supplementary feeding, while cautioning against disturbing the colony's thermoregulation when feeding them. It is a cheap, environmentally friendly solution to reducing bee stress.

Bonnefond et al. (2021) tested muzzles: the STOP-IT plastic muzzle and 25-mm chicken-wire curled around the whole of the front of the hive. They also tested 'ploys': dead hornets hanging from threads outside the hive entrance, intended to put off hornets from hawking. The ploys did not work; but they found that, again, the muzzle was effective in allowing bees to fly, even though the number of bees actually caught did not change significantly. The numbers of hornets hawking was not affected.

Whether the muzzle has a solid floor and/or sides is up for experimentation: will more exits for the bees help? Some beekeepers simply curl a piece of wire mesh from below the landing board up to the roof, and tuck it under the roof; and I have seen whole hives under nets — but that would be quite awkward to deal with when inspecting hives, unless the mesh tent is as roomy as a fruit cage.

Emmaüs Lescar-Pau (Emmaüs s a huge recycling charity in France) have produced plans for a bee house to protect bees from hornets. This bee house uses a fine mesh that hornets cannot get through, so keeps the hornets away from the hives, reducing the stress caused by them when they hunt in numbers in front of the hives. I don't know whether they become targets when negotiating the mesh.

Finally, another approach is the 'vertical muzzle', which in fact is a tunnel or 'chimney', 4 cm wide, that forces bees and hornets to fly down vertically to the hive entrance (Turchi & Derijard 2018), putting hornets into a vulnerable position where they could be attacked by guards. So far, this idea does not seem to have gained traction.

Reduction of entrances

A simple action to stop yellow-legged hornets getting inside your hive is to use a guard (similar to a mouse guard) with holes no bigger than 5.5 mm. Yellow-legged hornets usually do not attempt to get inside the hive unless it is small and weak and therefore relatively unguarded. Kevin Baughen says that beekeepers in his part of France (between Limoges and Poitiers) use plastic anti-hornet entrance reducers in the autumn when colony numbers are dwindling, to prevent hornets from entering weaker colonies. They then stay on as mouse guards for the winter. Drones do struggle to get through the gaps (which are 5 mm high), so the guards are removed before drone build-up in the spring. They use the French Nicot green plastic entrance reducers (see photo on p 66). Some YLHs are small enough to get through these guards, in which case a queen-excluder can be considered (see below).

Apiary management

Hand-in-hand with any kind of intervention to reduce hornets or disrupt hawking (unless you can stop predation completely by getting the nest destroyed) is honey bee colony management. If yellow-legged hornets become established, then when and how beekeepers look after their colonies will have to change. Such management adjustments can be divided into two approaches: (1) reducing the impact of yellow-legged hornets by choosing when to work the hives, and (2) preventative measures.

Choosing when to work

We don't yet know the exact timings of the yellow-legged hornet life cycle if it becomes established in the UK (and it will vary depending on weather and latitude), but we are likely to have a shorter season than in France. From the beginning of serious predation by the hornet in the apiary (July or, more likely, August), until the collapse of hornet colonies at the end of the year (October or November), opening hives is going to be difficult. Any manipulations, inspections, health interventions, feeding and honey removal will have to be moved to earlier in the year, or done at dusk or dawn before hornets are flying (and without stressing the bees by chilling them).

Preventative measures

Preventative measures are all about reducing stress and increasing the resilience of your bees. Diéguez-Antón et al. (2022b) found that the colonies that did best in the face of yellow-legged hornets were the ones where the whole colony was working properly, indicated by them keeping on top of hive thermoregulation. Conversely, hornet pressure was worst for small weak colonies with high varroa infestation rates, or those that had a combination of several risk factors.

Stress can be lessened by reducing contact with hornets, and the smell and sound of hornets, too, by using solid floors in the hive, or covering the mesh in an open-mesh floor (OMF). Using measures to reduce numbers of hornets (such as using electric harps or traps) and keeping the hornets at a distance from the hive entrance by using a muzzle, vegetation or pile of twiggy material really does help. Introducing such measures in a timely way is important so that the bees can get used to them, especially muzzles or Koldo traps.

When bees are threatened by YLHs, especially if small hornets can get through a hornet guard at the entrance to the hive, they tend to form a cluster around the queen and do fewer of the tasks that would have them roaming around the comb (feeding young, cleaning, removing dead bodies etc). This behaviour is generally bad for the colony, and it can be relieved by putting a queen excluder (this is a grille or slotted sheet that allows workers to pass through, but not queens or males, known as drones) between the bottom of the brood box and the entrance. With this in place, the colony can relax and revert to normal duties. It also allows for maximum contact with miticide varroa treatments across the combs, should this be necessary (Diéguez-Antón et al., 2024). There is a danger to this practice, though. As mentioned, drones cannot pass through queen excluders and so can't get out of the hive. They die trying to get through this barrier and their bodies clog the holes in the queen excluder, possibly leading to asphyxiation of the whole colony. Therefore queen excluders shouldn't be used at times when plenty of drones are around. Also, you will want a queen to be able to leave the hive to go on a mating flight, so use of queen excluders below the brood box should be limited to times of maximum hornet pressure in late summer/autumn, when queens will not be mating and after drones have been evicted.

If bees are unhealthy, and suffering from disease or varroa, they will succumb more quickly to the effects of hornet predation, so maintaining healthy bees and practicing good hygiene become even more important. Because varroa is such a widespread problem in beekeeping, and weakens colonies significantly, it needs to be dealt with well before hornet pressure builds in the late summer or autumn, either by miticide treatment, or by a brood break, such as performing a shook swarm.

Small colonies don't have foragers to spare, and should be united into bigger colonies as soon as possible. And you may want to be more effective with swarm control.

Finally, if colonies are suffering from foraging paralysis, they will need feeding. Again and again, reports from France emphasise the amount of extra feeding required to keep colonies alive. And don't forget that if your bees have stopped going out altogether (complete foraging paralysis), then they are not getting water, nectar, pollen, or resin for propolis. At a time when the winter bees are larvae, and need to be properly fed to result in strong and healthy adults to see the colony through to next spring, you must make sure that your bees are getting enough water and diverse nutrients and energy from pollen and syrup. Poor nutrition also leads to compromised immune function and increased susceptability to disease and agrochemicals in honey bees.

How much is this going to cost beekeepers?

In a paper that came out in 2018 (Ferreira-Golpe et al.), beekeepers and bee farmers in the province of Coruña, Spain, were asked about what they were doing to counteract yellow-legged hornets and how much these measures were costing them. The Spanish beekeepers used trapping most frequently, followed by reducing the hive entrance size, nest destruction and the 'Trojan' method (poisoned insect goes back to nest). Only six farms moved their hives (transhumance), three used electric harps and one used muzzles. There was also the cost of extra feeding to help colonies survive. All in all, they used 20% of the value of their production in combatting yellow-legged hornets.

Yellow-legged hornet with an attached streamer feeds at a wick-pot. Despite the wonky angle of the streamer, this hornet could fly well.

Glossary

abdomen
The bulbous back half of a hornet (see anatomy drawing)

AHAT
Asian Hornet Action Team

APHA
Animal and Plant Health Agency, an executive agency of Defra

apiary
A collection of hives managed by a beekeeper

BBKA
British Beekeepers' Association

brood
Young life stages of honey bees and hornets (eggs, larvae, pupae)

callow
A freshly emerged (eclosed) adult (see teneral)

capture-mark-recapture
A technique used by biologists to estimate the number of organisms in an area

castes
The distinct forms that occur in a colony of a social insect: queen, worker, male

cell
Hexagonal hole in comb in which a single larva/pupa is raised

colony
A collection of social insects (usually with a single queen) that act together in an organised way to raise offspring to produce the next generation

cocoon
A silk enclosure, within which the larva turns into an adult (pupation)

comb
The plates of hexagonal cells in which young are raised: they are horizontal and made of paper in wasps and vertical and made of wax in honey bees

cuticle
The skin of an insect — it may be thin and flexible, or thick where the exoskeleton needs to be strong.

Defra
Department for Environment, Food & Rural Affairs, a government department

diploid
Having two sets of each chromosome (twice that of haploid)

eclosion
Usually used by entomologists to describe the process of emerging from the pupa as an adult insect. Can also refer to a larva hatching from an egg

embryo nest
The initial YLH nest built solely by the foundress queen. No workers are present yet

endocrine
To do with the internal secretion of hormones

entomofauna
The insects found in a particular place

entomologist
Someone who studies insects

ESA
Epidémiosurveillance Santé Animale: French public biosecurity organisation

exoskeleton
The system of strong plates linked by flexible membranes which makes up the outer part and support structure of an adult insect, like a suit of armour

exudate
Something which exudes from something else: in this case some kind of liquid oozing out of a tree; a sap run

family
A rank or group used by taxonomists (those that work out the relatedness of organisms). It is above genus, which is above species. E.g. family = Vespidae, genus = *Vespa*, species = *velutina*)

FDGDON
La Fédération Départementale des Groupements de Défense contre les Organismes Nuisibles

Fera Science Ltd.
Formerly the Food and Environment Research Agency; now a joint venture between Capita and Defra

field, the
Experiments done in natural(ish) environments as opposed to in the laboratory

flight-mill
Device consisting of a small stand with a freely-rotating wire arm to which the insect is tethered. It is used to measure speed of flight and lengths of time the insect is willing to fly

foundress
YLH mated queen who has survived the winter and is now starting to establish a colony

FREDON
Fédération Régionale de Défense contre les Organismes Nuisibles

gaster
Another name for abdomen, used especially in the Hymenoptera

genus (pl. genera)
A group of closely related species, which all share the same first name, e.g. *Vespa* (it is always capitalized and should also be in italics)

gyne
A female sexual: once mated she will become a queen

haploid
Having a single set of unpaired chromosomes

Hymenoptera
An order of insects which includes bees, wasps, ants, sawflies etc.

IAS
Invasive alien species — category recognised by the European Union

INPN
Inventaire National du Patrimoine Naturel. Catalogue of natural biodiversity and geodiversity in France, managed by MNHN

INRA
L'Institut National de la Recherche Agronomique. French National Institute for Agricultural Research

instar
Stage of an insect between one moult and the next

larva, pl. larvae
In insects, it is the early immature stage, before pupation; in some insects they are known as maggots, grubs or caterpillars

mandibles
Pair of hard mouthparts used for biting, cutting or holding food. Also used in nest building in hornets

meconium (pl. meconia)
Waste from the gut expelled by hornet larva before it pupates

MNHN
Museum National d'Histoire Naturelle: French National Museum of Natural History

moult
Shedding of larval skin once it has become too tight; a new one lies beneath with more room for expansion

n
In academic biology articles, 'n' stands for number, so if n=9, then the data is from 9 individuals

NBU
National Bee Unit: part of APHA

nest
In hornets, the paper shelter built by the workers to house adults and young

NNSS
GB Non-native Species Secretariat

olfaction
The sense of smell or the act of smelling

petiole
The waist of the hornet, between the thorax and abdomen (see anatomical drawing opposite). Also the stalk which connects the combs to the supporting structure above them (see structures of primary and secondary nests)

pheromone
A secreted or excreted chemical factor that triggers a social response in members of the same species

primary nest
Once the embryo nest contains workers it is called a primary nest. Some authors do not recognise embryo nests and simply talk about primary or secondary nests

propolis
A red-brown, sticky, resinous material made by honey bees from material gathered from tree buds. It is used by the bees to fill gaps in hives among many other uses

pupa (pl. pupae)
Stage in the hornet between larva and adult, during which a complete transformation takes place from the larval form to the adult form. The change takes place hidden in the cocoon

queen
The sole reproductive female in a colony of hornets or honey bees. Also referred to as 'gyne' before mating, as 'foundress' when starting a new colony in the spring and

'mother queen' in a mature colony, to distinguish her from the young gynes which may also be present

RFID
Radio frequency identification, a technology which uses tiny computer chips to track individual insects

secondary nest
If a colony of hornets relocates, the nest which is built is called the secondary nest

sexuals
Males (called 'drones' in honey bee terminology, but not in hornet terminology) and females which will mate with males from other colonies. The female sexuals are called gynes in hornets and are destined to become queens

species
A type of organism. In *Vespa velutina nigrithorax*, '*Vespa*' is the genus name, '*velutina*' is the species name, and '*nigrithorax*' is the subspecies or colour form. The species and subspecies names are never capitalized, but the whole name is usually in italics

tegula (pl. tegulae)
One of a pair of small hooded covers over the join between the wing and the thorax

teneral
Recently emerged adult (see callow)

thorax
Part of the adult insect between the head and the abdomen. It houses powerful flight muscles which work the wings. The legs are also attached to the thorax

trophallaxis
The transfer of regurgitated liquid food between adults or adults and larvae

UAV
Unmanned Aerial Vehicle, commonly called a 'drone'

UKCEH
Centre for Ecology and Hydrology

worker
A usually non-reproducing female in a colony of hymenopteran insects. She will perform a range of different jobs (sometimes according to age) such as forager, guard, builder, cleaner, carer for larvae

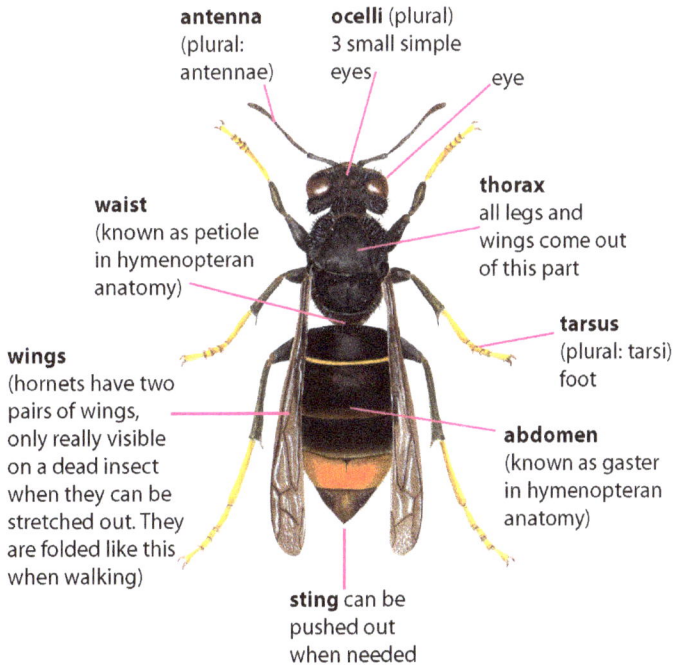

antenna (plural: antennae)

ocelli (plural) 3 small simple eyes

eye

waist (known as petiole in hymenopteran anatomy)

thorax all legs and wings come out of this part

wings (hornets have two pairs of wings, only really visible on a dead insect when they can be stretched out. They are folded like this when walking)

tarsus (plural: tarsi) foot

abdomen (known as gaster in hymenopteran anatomy)

sting can be pushed out when needed

BOX 7. How to make a hands-free streamer pot
Technique developed by Chris Isaacs

You will need:

Queen bee catcher with 36 mm inside diameter.

Plastic tube or other cylindrical object with 35 mm outside diameter, 70 mm tall. Alkathene pipe of the right size is another option, but needs to be cut square so that the plunger fits against the mesh evenly.

Pair of side-cutters, scissors, craft knife, heavy-duty scissors or tin-snips.

Electrical insulation tape.

Piece of 2-mm-thick lead flashing.

Optional for base: 25mm by 8 mm ferrite magnet, glue, metal jam jar lid.

Instructions:

Use the side-cutters to remove some mesh in the queen catcher lid to create a slot. The mesh is brittle, so take care. Scrape off the rough edges using a craft knife or a nail-file.

Remove the wooden stem from the queen catcher plunger. If it doesn't pull off, break it off and trim any splinters sticking out. Attach the old sponge and wooden plunger top to the base of the new plastic tube with electrical insulation tape.

Use heavy-duty scissors or tin snips to cut a (roughly) 13-mm-wide strip from 2-mm-thick lead flashing as a weight to keep the hornet still. Use gloves: lead is toxic. Wrap the lead strip around the queen catcher, near the open end, and trim to length. Tape in position with electrical tape. Make sure the lead is covered by tape to avoid later hand contact.

Pop the plunger into the queen catcher and check that it moves freely.

When using, the tube is more stable without its screw-on lid, but a magnet can be glued into the screw-on lid. Then, when the lid is screwed on, it will stick to a steel plate such as a jam-jar lid, making a very stable base. The magnet enables the base to be removed while catching hornets or storing the pot in a pocket.

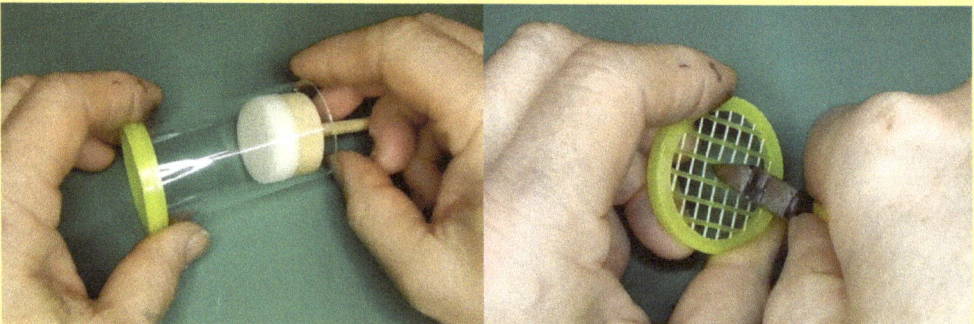

1. A queen catcher.

2. Cut slot in grid using side-cutters.

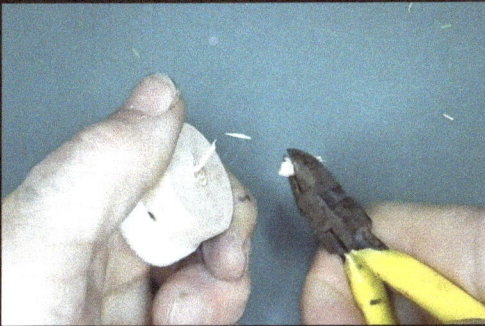

3. Remove wooden stem.

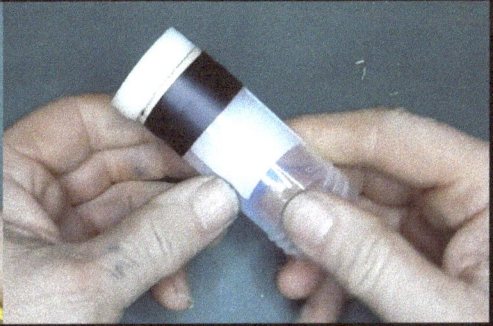

4. Tape sponge+wood to base of tube.

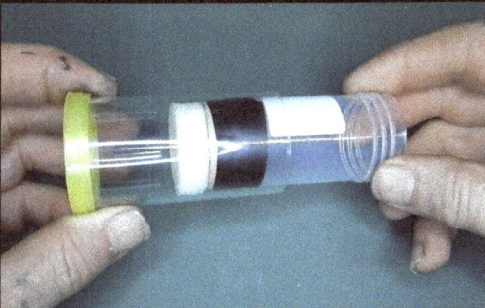

5. Make sure new plunger slides freely.

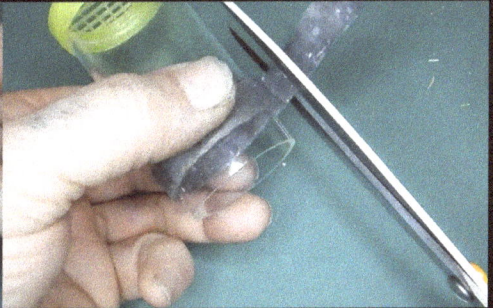

6. Trim lead strip to length.

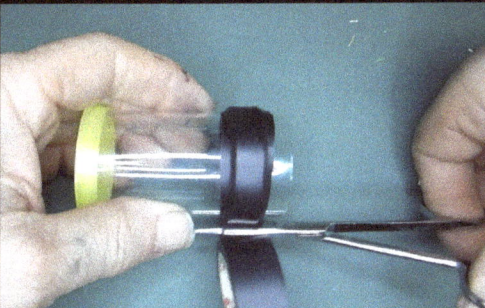

7. Tape lead onto queen catcher.

8. Glue ferrite magnet into screw-lid.

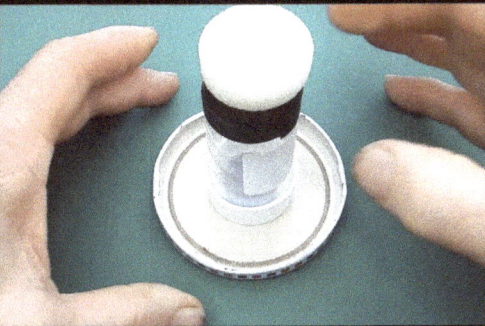

9. Jam-jar lid makes a stable base.

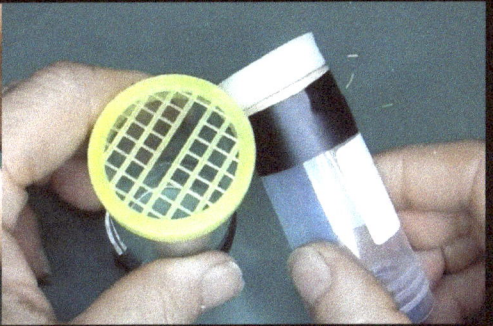

10. Finished streamer pot.

Resources

13 Bees Website of Kevin and Amanda Baughen - Bee friendly holidays in France www.13bees.co.uk

AHATs Asian Hornet Action Teams www.ahat.org.uk/

BBKA British Beekeepers' Association www.bbka.org

BBwear www.bbwear.co.uk/

BWARS Bees, Wasps & Ants Recording Society www.bwars.com

Electric harp - Spanish manufacturers - www.sanve.weebly.com

FDGDON du Morbihan La Fédération Départementale des Groupements de Défense contre les Organismes Nuisibles: http://www.fredon-bretagne.com/fdgdon-morbihan/

Fit2Fight, Practical Aid for Beekeepers in Managing the Asian Hornet by Alan Baxter (available from Northern Bee Books)

FREDON (Bretagne) Fédération Régionale de Défense contre les Organismes Nuisibles www.fredon-bretagne.com

Jersey Asian Hornet Group facebook page https://www.facebook.com/groups/293640477963172/ see also 'Jersey Asian hornet diary' on YouTube - superb videos from John de Carteret

LOYDERN Spanish company making system for attaching a small capsule to worker. Used for poisoning nest. Worth trying with diatomaceous earth? www.loydern.com In Spanish.

Muzzle, plastic. STOP-IT, www.stop-it.fr

MNHN Muséum National D'Histoire Naturelle Up-to-date map of European spread of Asian hornets http://frelonasiatique.mnhn.fr/home/

NBU National Bee Unit http://www.nationalbeeunit.com

NNSS GB Non-native Secretariat http://www.nonnativespecies.org

Northern Bee Books Specialist booksellers on everything to do with bees www.northernbeebooks.co.uk

Paintball Asian hornet insecticide gun www.frelons.com

Robor Nature Radio-tracking kits for finding YLH nests. Inspired by Exeter University's pioneering work. www.robor-nature.eu (can be read in English).

UNAF Union National De L'Apiculture Française www.unaf-apiculture.info/

uni POSCA marker pens their website: www.posca.com, wide range from www.cultpens.com

AHAT kit list

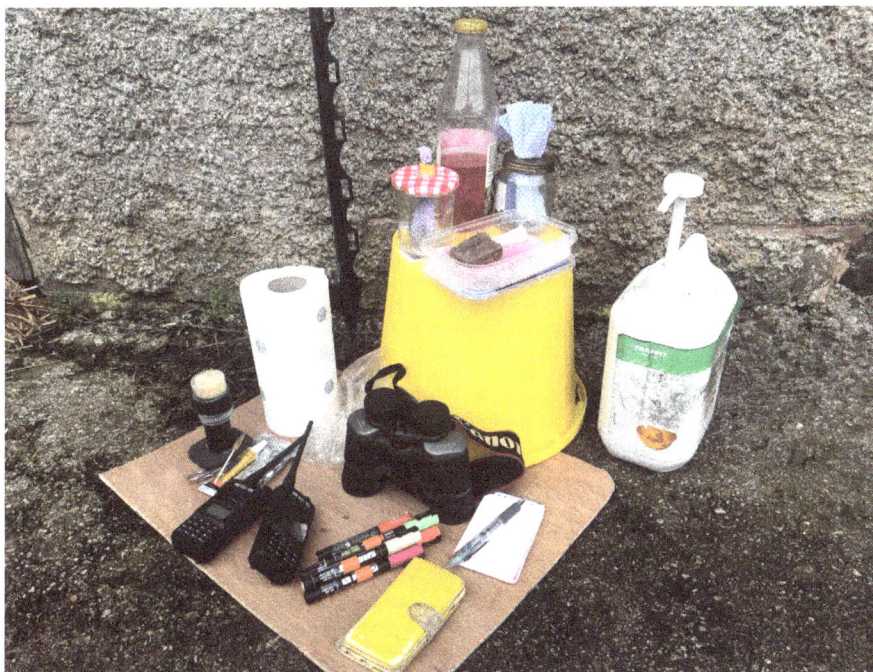

Dish-type bait stations, plastic containers with lids are ideal (e.g. takeaway boxes).

Wick-pot bait stations.

Plastic electric-fence post is useful for hanging a wick-pot on (the plastic ones have wide tops, so you are less likely to poke your eye).

Bait - e.g. Trappit or homemade. Make sure it is in a labelled container.

Streamer kit, including a **queen catcher.**

Paper **kitchen roll** or similar.

Smartphone, otherwise a stopwatch and compass.

uni POSCA marker pens (PC-3M), several colours (light colours are best).

Small **notebook** and **pen.**

Yellow builders' bucket.

Walkie-talkies (make sure different AHAT teams use compatible models so that if teams work together, they can listen AND talk to one another). If you buy ones with integral rechargable batteries and these are not kept charged over the winter, you will ruin the batteries, and possibly the walkie-talkie. If you use removable rechargable batteries which are removed over the winter you can avoid this. **Rechargeable batteries** (plenty) plus charger (if that's what your walkie-talkies use).

Binoculars (usually a personal item, but may need to buy).

Copy of this **book**!

Sturdy container to keep everything in.

References

AAAFAa (2016) Association Action Anti Frelon Asiatique. The jar-and-card technique (video). https://anti-frelon-asiatique.com/piegeage/detruire-un-jeune-nid-sans-insecticides-et-pour-100-mag-m6/ (Accessed 6 March 2019)

AAAFAb (2015) Association Action Anti Frelon Asiatique. Sieves on a litter-picker technique (video). https://anti-frelon-asiatique.com/piegeage/recuperation-100-naturelle-dun-nid-primaire/ (Accessed 6 March 2019)

AAAFAc (2016) Association Action Anti Frelon Asiatique. Capture Reine dans Camélia - Pince à passoires. https://vimeo.com/156297742 (Accessed 19 April 2019)

Arca M, Papachristoforou A, Mougel F, Rortais A, Monceau K, Bonnard O, Tardy P, Thiéry D, Silvain J-F, Arnold G (2014) Defensive behaviour of *Apis mellifera* against *Vespa velutina* in France: testing whether European honeybees can develop an effective collective defence against a new predator. Behavioural Processes 106, 122-129

Arca M, Mougel F, Guillemaud T, Dupas S, Rome Q, Perrard A, Muller F, Fossund A, Capdevielledula C, Torres-Leguizamon M, Chen XX, Tn JL, Jung C, Villemant C, Arnold G, Silvain JF, (2015) Reconstructing the invasion and the demographic history of the yellow-legged hornet, *Vespa velutina*, in Europe. Biol. Invasions, 17: 2357-2371

Archer ME (2008) Taxonomy, distribution and nesting biology of species of the genera *Provespa* Ashmead and *Vespa* Linnaeus (Hymenoptera, Vespidae). Entomologist''s Monthly Magazine. ;144(1727):69.

Archer, ME (1985) Population dynamics of the social wasps *Vespula vulgaris* and *Vespula germanica* in England. Journal of Animal Ecology, 54, 473 – 485.

Balmori A (2015) Sobre el riesgo real de una expansión generalizada de la avispa asiática *Vespa velutina* Lepeletier, 1836 (Hymenoptera: Vespidae) en la Península Ibérica. Boletín de la SEA. (56):283-9. (In Spanish)

Barbet-Massin M, Rome Q, Muller F, Perrard A, Villemant C, Jiguet F (2013) Climate change increases the risk of invasion by the yellow-legged hornet. Biol Conserv 157:4-10

Barbet-Massin M, Salles JM, Courchamp F. (2020) The economic cost of control of the invasive yellow-legged Asian hornet. NeoBiota. 2020;55:11-25.

Barracchi D, Cusseau G, Pradella D, Turillazzi S (2010) Defence reactions of *Apis mellifera ligustica* against attacks from the European hornet *Vespa crabro*. Ethology Ecology & Evolution 22: 281-294

Barandika JF, de la Hera O, Fañanás R, Rivas A, Arroyo E, Alonso RM, Alonso ML, Galartza E, Cevidanes A, García-Pérez AL. (2023) Efficacy of Protein Baits with Fipronil to Control *Vespa velutina nigrithorax* (Lepeletier, 1836) in Apiaries. Animals. Jun 23;13(13):2075.

Beggs JR, Brockerhoff EG, Corley JC, Kenis M, Masciocchi M, Muller F, Rome Q, Villemant C (2011) Ecological effects and management of invasive Vespidae. Biocontrol 56: 505–526

Bertolino S, Lioy S, Laurino D, Manino A, Porporato M (2016) Spread of the invasive yellow-legged hornet *Vespa velutina* (Hymenoptera: Vespidae) in Italy. Appl Entomol Zool 51: 589-597

Bob Binnie (2023) Yellow Legged Hornet. YouTube video https://www.youtube.com/watch?v=4qNJuZw56KA. Accessed December 2023.

Bonnefond L, Paute S, Andalo C. (2021) Testing muzzle and ploy devices to reduce predation of bees by Asian hornets. Journal of Applied Entomology. Feb;145(1-2):145-57.

References

Bouchebti S, Bodner L, Bergman M, Magory Cohen T, Levin E. (2022) The effects of dietary proline, beta-alanine, and gamma-aminobutyric acid (GABA) on the nest construction behavior in the Oriental hornet (Vespa orientalis). Scientific Reports. May 6;12(1):7449.

Blanchard P, Schurr F, Celle O, Cougoule N, Drajnudel P, Thiéry R, Faucon J-P, Ribiére M (2008) First detection of Israeli acute paralysis virus (IAPV) in France, a dicistrovirus affecting honeybees (*Apis mellifera*). J. Invertebr. Pathol. 99: 348-350

Blot J (2007) Le frelon asiatique (*Vespa velutina*). Le piégeage des fondatrices. Fiche technique apicole. Bull. Tech. Apic. 34 (4) 201-204 (in French)

Budge GE, Hodgetts J, Jones EP, Ostojá-Starzewski JC, Hall J, Tomkies V, Semmence N, Brown M, Wakefield M, Stainton K (2017) The invasion, provenance and diversity of *Vespa velutina* Lepeletier (Hymenoptera: Vespidae) in Great Britain. PLoS ONE 12(9)

Cappa F, Cini A, Pepiciello I, Petrocelli I, Inghilesi AF, Anfora G, Dani FR, Bortolotti L, Wen P, Cervo R (2019) Female volatiles as sex attractants in the invasive population of *Vespa velutina nigrithorax*, Journal of Insect Physiology 119: 103952, ISSN 0022-1910 https://doi.org/10.1016/j.jinsphys.2019.103952.

Cappa F, Cini A, Meriggi N, Poidatz J, Thiery D, Cervo R (2022) Immune competence of the invasive hornet *Vespa velutina* and its native counterpart *Vespa crabro*: a comparison across caste and sex. Entomologia Generalis 42: 11-20

Caragata CR & Montesinos JLV (2020) Datos ambientales preliminares del avispón asiático (*Vespa velutina* Lepeletier, 1836) (Hymenoptera, Vespidae) en Asturias, España. Bol. R. Soc. Esp. Hist. Nat. 114 (in Spanish with English abstract)

Carisio L, Cerri J, Lioy S, Bianchi E, Bertolino S, Porporato M (2022) Impacts of the invasive hornet *Vespa velutina* on native wasp species: a first effort to understand population-level effects in an invaded area of Europe. Journal of Insect Conservation 26: 663-671

Carriazo S and Ortiz A (2022). Wasp stings and plasma exchange. Clinical kidney journal, 15(8), pp.1455-1458.

Carpenter JM, Kojima J (1997) Checklist of the species in the subfamily Vespinae (Insecta: Hymenoptera: Vespidae) Natural History Bulletin of Ibaraki University 1: 51-92

Carvalho J, Hipólito D, Santarém F, Martins R, Gomes A, Carmo P, Rodrigues R, Grosso-Silva J, Fonseca C. (2020) Patterns of *Vespa velutina* invasion in Portugal using crowdsourced data. Insect Conservation and Diversity. Sep;13(5):501-7.

Chauzat M, Martin S (2009). A foreigner in France, biological information on the Asian hornet *Vespa velutina*, a recently introduced species. Biologist 56: 86-91

Cheng Y, Wen P, Dong S, Tan K, Nieh JC (2017) Poison and alarm: the Asian hornet uses sting venom volatiles as an alarm pheromone. Journal of Experimental Biology 220: 645-651

Cheng YN, Wen P, Tan K, Darrouzet E. (2022) Designing a sex pheromone blend for attracting the yellow-legged hornet (*Vespa velutina*), a pest in its native and invasive ranges worldwide. Entomologia Generalis.

Chibrac J (2013) Un drone lutte contre les frelons asiatiques. Video. https://www.youtube.com/watch?v=KTkQ2fJfXyU

Choi MB, Martin SJ, Lee JW (2012) Distribution, spread, and impact of the invasive hornet *Vespa velutina* in South Korea, Journal of Asia-Pacific Entomology, 15 (3): 473-477

Choi M (2021a) Foraging behavior of an invasive alien hornet (*Vespa velutina*) at *Apis mellifera* hives in Korea: Foraging duration and success rate. Entomological Research 51: 143–148

Choi MB, Hong EJ, Kwon O. (2021) Defensive behavior of the invasive alien hornet, *Vespa velutina*, against color, hair and auditory stimuli of potential aggressors. PeerJ. 2021 Apr 6;9:e11249.

Choi MB (2021b) Defensive behavior of the invasive alien hornet *Vespa velutina nigrithorax* against potential human aggressors. Entomological Research. 2021 Apr;51(4):186-95.

Cini A, Cappa F, Petrocelli I, Pepiciello I, Bortolotti L, Cervo R (2018) Competition between the native and the introduced hornets *Vespa crabro* and *Vespa velutina*: a comparison of potentially relevant life-history traits. Ecological Entomology 43 (3) 351-362

Cini A, Meriggi N, Bacci G, Cappa F, Vitali F, Cavalieri D, Cervo R (2020) Gut microbial composition in different castes and developmental stages of the invasive hornet *Vespa velutina nigrithorax*. Science of The Total Environment 745: 140873, ISSN 0048-9697,https://doi.org/10.1016/j.scitotenv.2020.140873.

Couto A, Monceau K, Bonnard O, Thiéry D, Sandoz J-C (2014) Olfactory attraction of the hornet *Vespa velutina* to honeybee colony odors and pheromones. PLoS ONE 9 (12)

Couto A, Mitra A, Thiéry D, Marion-Poll F and Sandoz J-C (2017) Hornets Have It: A Conserved Olfactory Subsystem for Social Recognition in Hymenoptera? Front. Neuroanat. 11:48. doi: 10.3389/fnana.2017.00048

Crespo N, Louzada J, Fernandes LS, Tavares PB, Aranha J (2022) Microscopic Identification of Anatomical Elements and Chemical Analysis of Secondary Nests of *Vespa velutina nigrithorax* du Buyson. Insects, 13, 537. https://doi.org/10.3390/insects13060537

Dalmon A, Gayral P, Decante D, Klopp C, Bigot D et al. (2019) Viruses in the Invasive Hornet *Vespa velutina*. Viruses, 11, 1041; doi:10.3390/v11111041

Darrouzet E, Gévar J, Dupont S (2014) A scientific note about a parasitoid that can parasitize the yellow-legged hornet, *Vespa velutina nigrithorax*, in Europe. Apidologie 46: 130-132

Darrouzet E, Gévar J, Guignard Q, Aron S (2015) Production of early diploid males by European colonies of the invasive hornet *Vespa velutina nigrithorax*. PLoS ONE 10(9): e0136680

Decante, D (2015) Lutte contre le frelon asiatique *Vespa velutina*: Évaluation comparative des modalités de piégeage de protection du rucher. http://itsap.asso.fr/wp-content/uploads/2016/03/cr_ evaluation) piegeage_vvelutina)2014.pdf (Arrives in downloads as a pdf. Accessed 18 February 2019) (in French)

Diéguez-Antón A, Escuredo O, Seijo MC, Rodríguez-Flores MS (2022a) Embryo, Relocation and Secondary Nests of the Invasive Species *Vespa velutina* in Galicia (NW Spain). Animals, 12, 2781. https://doi.org/10.3390/ani12202781

Diéguez-Antón A, Rodríguez-Flores MS, Escuredo O, Seijo MC (2022b) Monitoring Study in Honeybee Colonies Stressed by the Invasive Hornet *Vespa velutina*. Vet. Sci. 2022, 9, 183. https://doi.org/10.3390/vetsci9040183

Diéguez-Antón A (2023). Estudio del comportamiento de la especie invasora *Vespa velutina nigrithorax* y de la presión ejercida sobre la especie *Apis mellifera iberiensis* en los colmenares gallegos (Doctoral dissertation, Bioloxía vexetal e ciencias do solo, Universidade de Vigo).http://hdl.handle.net/11093/6279

References

Diéguez-Antón A, Rodríguez-Flores MS, Ordoñez Dios ME, Bunker S, Nave A, Godinho J, Casaca JD, Guedes H, Seijo MC (2024) Management of honey bee (Hymenoptera: Apidae) colonies under yellow-legged hornet (Hymenoptera: Vespidae) pressure. Journal of Integrated Pest Management. 2024;15(1):43

Do Y, Park WB, Park JK, Kim CJ, Choi MB. (2022) Genetic and morphological variation of *Vespa velutina nigrithorax* which is an invasive species in a mountainous area. Scientific Reports. Mar 18;12(1):4737.

Dong D, Wang Y (1989) A preliminary study on the biology of wasps *Vespa velutina auraria* Smith and *Vespa tropica ducalis* Smith (Hymenoptera: Vespidae). Zoological Research 10(2): 155-162

Dong S, Wen P, Zhang Q, Wang Y, Cheng Y et al. (2018) Olfactory eavesdropping of predator alarm pheromone by sympatric but not allopatric prey. Animal Behaviour 141: 115e125

Dong S, Tan K, Zhang Q, Nieh JC (2019) Playbacks of Asian honey bee stop signals demonstrate referential inhibitory communication. Animal Behaviour 148: 29e37

Dong S, Gu G, Li J, Wang Z, Tan K, Yang M, Nieh JC (2023) Honey bee social collapse arising from hornet attacks. DOI: 10.1127/entomologia/2023/1825

Edwards R (1980) Social wasps: their biology and control, The Rentokil Library, East Grinstead

Edwards E, Toft R, Joice N, Westbrooke I (2017) The efficacy of Vespex® wasp bait to control *Vespula* species (Hymenoptera: Vespidae) in New Zealand. International Journal of Pest Management 63: 266-272

FDGDON Bretagne (2024) https://www.gie-elevages-bretagne.fr/admin/upload/20240416_bilan_pression_frelon_asiatique_2023_Bretagne.pdf (accessed 27 February 2025; in French)

FGDON35 (2025) https://fgdon35.fr/nos-missions/frelon-asiatique/ (accessed 27 February 2025; in French)

Feás X (2021) Human Fatalities Caused by Hornet, Wasp and Bee Stings in Spain: Epidemiology at State and Sub-State Level from 1999 to 2018. Biology 10, 73. https://doi.org/10.3390/ biology10020073

Feás X, Vidal C, Vázquez-Tato MP, Seijas JA (2022) Asian Hornet, *Vespa velutina* Lepeletier 1836 (Hym.: Vespidae), Venom Obtention Based on an Electric Stimulation Protocol. Molecules 27, 138. https://doi.org/10.3390/molecules27010138

Ferreira-Golpe MA, García Arias AI, Pérez-Fra M (2018) Costes de la lucha contra la especie invasora *Vespa velutina* soportados por los apicultores en la provincia de a Coruña. Conference paper, CIER XII, Segovia 4th, 5th and 6th July 2018. (in Spanish)

Ferreira-Golpe MAF, Arias AIG and González IV (2019). Propuesta metodológica para determinar la influencia de la presencia de la "*Vespa velutina*" en la variación de la producción. In La sostenibilidad agro-territorial desde la Europa atlántica (pp. 605-608). Asociación Española de Economía Agraria, AEEA. (In Spanish).

Franklin DN, Brown MA Datta S, Cuthbertson AGS, Budge GE, Keeling MJ (2017) Invasion dynamics of Asian hornet, *Vespa velutina* (Hymenoptera: Vespidae): a case study of a commune in south-west France. Appl Entomol Zool 52: 221

Gabín-García LB, Bartolomé C, Guerra-Tort C, Rojas-Nossa SV et al. (2021) Identification of pathogens in the invasive hornet *Vespa velutina* and in native Hymenoptera (Apidae, Vespidae) from SW-Europe. Nature Scientific Reports 11: 11233

García-Arias AI, Ferreira-Golpe MA, Vázquez-González I, et al. (2023) Economic costs and practices to control *Vespa velutina nigrithorax* in beekeeping: a survey in four regions in Europe. Training (TRA), 40, pp.8-3.

Garnas R, Auger-Rozenberg J, Roques MA, et al. (2016) Complex patterns of global spread in invasive insects: eco-evolutionary and management consequences. Biol Invasions 18, 935–952 (2016). https://doi.org/10.1007/s10530-016-1082-9

Gévar J, Bagnères A-G, Christidès J-P, Darrouzet E (2017) Chemical heterogeneity in inbred European population of the invasive hornet *Vespa velutina nigrithorax*. J Chem Ecol 42: 763-777

Gibo DL, Temporale A, Lamarre TP, Soutar BM, Dew HE (1977) Thermoregulation in colonies of *Vespa arenaria* and *Vespula maculata* (Hymenoptera: Vespidae) III. Heat production in queen nests. Can Entomol 109:615-620

Godinho J, Santos M, Nave A (2020) A *Vespa velutina* e os desafios para a apicultura na região Norte de Portugal – contributos do projeto Gesvespa. Agrociência Junho 2020 Voz Do Campo (In Portuguese)

Gourbière S, Menu F (2009). Adaptive dynamics of dormancy duration variability: evolutionary trade-off and priority effect lead to suboptimal adaptation. Evolution: International Journal of Organic Evolution. 63. 1879-92.

Greene A (1999) *Dolichovespula* and *Vespula*. In: Ross KG, Matthews RW, Eds. The Social Biology of Wasps. Ithaca, New York: Cornell University Press 1999; 263-305.

Gu G, Meng Y, Tan K, Dong S, Nieh JC (2021) Lethality of Honey bee Stings to Heavy Armored Hornets. Biology. 10, 484. https://doi.org/10.3390/biology10060484

de Haro L, Labadie M, Chanseau P, Cabot C, Blanc-Brisset I, Penouil F, National Coordination Committee for Toxicovigilance (2010) Medical consequences of the Asian black hornet (*Vespa velutina*) invasion in Southwestern France. Toxicon 55: 650-652

Hassall RM, Purse BV, Barwell L, Booy O, Lioy S, Rorke S, Smith K, Scalera R, Roy HE. (2025) Predicting the spatio-temporal dynamics of biological invasions: Have rapid responses in Europe limited the spread of the yellow-legged hornet (*Vespa velutina nigrithorax*)?. Journal of Applied Ecology. 2025 Jan;62(1):106-18.

Haxaire J, Bouguet J-P and Tamisier J-P (2006) *Vespa velutina* Lepeletier, 1836, une redoutable nouveauté pour la faune de France (Hym., Vespidae). Bulletin de la Société entomologique de France 111: 194 (in French)

Haxaire J , Villemant C (2010) Impact sur l'entomofaune des "piéges à frelon asiatique". Insectes no. 159 (4) (in French)

de la Hera O, Alonso ML, Alonso RM (2023) Behaviour of *Vespa velutina nigrithorax* (Hymenoptera: Vespidae) under Controlled Environmental Conditions. Insects, 14, 59. https://doi.org/10.3390/insects14010059

Herrera C, Marqués A, Colomar V, Leza, MM (2019). Analysis of the secondary nest of the yellow-legged hornet found in the Balearic Islands reveals its high adaptability to Mediterranean isolated ecosystems. In: Veitch, C.R., Clout, M.N., Martin, A.R., Russell, J.C., West, C.J. (Eds.), Island Invasives: Scaling up to Meet the Challenge. IUCN, Gland, pp. 375–380.

Herrera C, Jurado-Rivera JA, Leza M. (2023) Ensemble of small models as a tool for alien invasive species management planning: Evaluation of *Vespa Velutina* (Hymenoptera: Vespidae) under Mediterranean island conditions. Journal of Pest Science. Jan;96(1):359-71.

Ikegami M, Tsujii K, Ishizuka A, Nakagawa N, Kishi S, Sakamoto Y, Sakamoto H, Goka K. (2020) Environments, spatial structures, and species competitions: determining the impact of yellow-legged hornets, *Vespa velutina*, on native wasps and bees on Tsushima Island, Japan. Biological

Invasions. Oct;22:3131-43.

Ishay J and Ikan R (1968) Food exchange between adults and larvae in *Vespa orientalis* F. Anim. Behav. 16: 298-303

Jeanne RL & Suryanarayanan S (2011) A new model for caste development in social wasps, Communicative & Integrative Biology, 4:4, 373-377, DOI: 10.4161/cib.15262

Jeon MS, Jeong Y, Lee J, Yu SH, Kim SB, Kim D, Kim KC, Lee S, Lee CW, Choi I. (2023) Deep Learning-Based Portable Image Analysis System for Real-Time Detection of *Vespa velutina*. Applied Sciences. Jun 22;13(13):7414.

Jeong H, Kim JM, Kim B, Nam JO, Hahn D, Choi, MB (2020) Nutritional value of the larvae of the alien invasive wasp *Vespa velutina nigrithorax* and amino acid composition of the larval saliva. Foods 9(7):885

Jeong Y, Jeon MS, Lee J, Yu SH, Kim SB, Kim D, Kim KC, Lee S, Lee CW, Choi I. (2023) Development of a Real-Time *Vespa velutina* Nest Detection and Notification System Using Artificial Intelligence in Drones. Drones. 2023 Oct 10;7(10):630.

Jones CJ and Oldroyd BP (2006) Nest thermoregulation in social insects. Advances in Insect Physiology 33 DOI: 10.1016/S0065-2806(06)33003-2

Jones EP, Conyers C, Tomkies V et al. (2020) Managing incursions of *Vespa velutina nigrithorax* in the UK: an emerging threat to apiculture. Sci Rep 10, 19553. https://doi.org/10.1038/s41598-020-76690-2

Ju C, Son HI (2022) Investigation of an autonomous tracking system for localization of radio-tagged flying insects IEEE Access. Jan 5;10:4048-62.

Jung C (2012) Spatial expansion of an invasive hornet, *Vespa velutina nigrithorax* Buysson (Hymenoptera: Vespidae) in Korea. Korean Journal of Apiculture 27, 87-93

Kanzaki N, Kosaka H, Sayama K, Takahashi J, Makino S (2007) *Sphaerularia vespae* sp. nov. (Nematoda, Tylenchomorpha, Sphaerularioidae), an endoparasite of a common Japanese hornet, *Vespa simillima* Smith (Insecta, Hymenoptera, Vespidae). Zoological Science, 24(11), 1134-1142. https://doi.org/10.2108/zsj.24.1134

Kanzaki N, Makino SI, Kosaka H, Sayama K, Hamaguchi K, Narayama S (2023) Nematode and Strepsipteran Parasitism in Bait-Trapped and Hand-Collected Hornets (Hymenoptera, Vespidae, Vespa). Insects. Apr 20;14(4):398.

Keeling MJ, Franklin DN, Datta S, Brown MA, Budge GE (2017) Predicting the spread of the Asian hornet (*Vespa velutina*) following its incursion into Great Britain. Scientific Reports (7) 6240

Kennedy PJ, Ford SM, Poidatz J, Thiéry D, Osborne JL (2018) Searching for nests of the invasive Asian hornet (*Vespa velutina*) using radio-telemetry. Communications Biology 1 (88)

Kennedy PJ & Osborne JL (2023a) A review of the success of the UK strategy to tackle the invasive insect *Vespa velutina nigrithorax*, the "Asian hornet". University of Exeter Open Research (ORE) identifier: http://hdl.handle.net/10871/133510.

Kennedy & Osborne, (2023b). A guide to radio-tracking Asian hornets to locate their nests. http://hdl.handle.net/10871/133513

Kim D, Jung C. (2019) Study on defensive behavioral mechanism of *Vespa* hornets by anthropogenic external disturbance. Journal of Apiculture. 2019 Sep 30;34(3):181-7. (In Korean with English abstract)

Kim C-J, Choi MB (2021) First Discovery of *Vespa velutina nigrithorax* du Buysson (Hymenoptera: Vespidae), an Invasive Hornet in the Feces of the Yellow-Throated Marten in South Korea. Insects, 12, 296.https://doi.org/10.3390/insects12040296

Kim B, Ju C, Son HI (2022) Field evaluation of UAV-based tracking method for localization of small insects. Entomological Research. Mar;52(3):135-47.

Kishi S, Goka K (2017) Review of the invasive yellow-legged hornet, *Vespa velutina nigrithorax* (Hymenoptera: Vespidae), in Japan and its possible chemical control. Appl Entomol Zool 52: 361-368

Klingner R, Richter K, Schmolz E, Keller B (2005) The role of moisture in the nest thermoregulation of social wasps. Naturwissenschaften 92: 427-430

Kwon O, Choi MB (2020) Interspecific hierarchies from aggressiveness and body size among the invasive alien hornet, *Vespa velutina nigrithorax*, and five native hornets in South Korea. PLoS ONE 15(7): e0226934. https://doi.org/10.1371/journal.pone.0226934

Laborde-Castérot H, Darrouzet E, Le Roux G, Labadie et al. (2021). Ocular lesions other than stings following yellow-legged hornet (*Vespa velutina nigrithorax*) projections, as reported to French Poison control centers. JAMA ophthalmology, 139(1), pp.105-108.

Laurino D, Lioy S, Carisio L, Manino A, Porporato M (2020) *Vespa velutina*: An Alien Driver of Honey Bee Colony Losses. Diversity 12: 5; doi:10.3390/d12010005

Lee CG, Yu SH. (2023) Exterminator for the Nests of *Vespa velutina nigrithorax* Using an Unmanned Aerial Vehicle. Drones. Apr 21;7(4):281.

Leza M, Herrera C, Marques A, Roca P, Sastre-Serra J, Pons DG (2019) The impact of the invasive species *Vespa velutina* on honeybees: A new approach based on oxidative stress. Science of the Total Environment 689 (2019) 709–715

Leza M, Herrera C, Picó G, Morro T, Colomar V. (2021) Six years of controlling the invasive species *Vespa velutina* in a Mediterranean island: The promising results of an eradication plan. Pest Management Science. May;77(5):2375-84.

Lim DJ, Lee JE, Lee JS, Kim I, Kim IS. (2019) Attraction of the Invasive Hornet, *Vespa velutina nigrithorax*, by using *Bacillus* sp. BV-1 Cultures. Korean Journal of Environmental Agriculture. ;38(2):104-9.

Lioy S, Manino A, Porporato M, Laurino D, Romano A, Capello M, Bertolino S (2019) Establishing surveillance areas for tackling the invasion of *Vespa velutina* in outbreaks and over the border of its expanding range.NeoBiota 46: 51-69. https://doi.org/10.3897/neobiota.46.33099

Lioy S, Laurino D, Maggiora R et al. (2021a) Tracking the invasive hornet *Vespa velutina* in complex environments by means of a harmonic radar. Sci Rep 11, 12143. https://doi.org/10.1038/s41598-021-91541-4

Lioy S, Bianchi E, Biglia A, Bessone M, Laurino D and Porporato M (2021b) Viability of thermal imaging in detecting nests of the invasive hornet *Vespa velutina*. Insect Science, 28: 271-277. https://doi.org/10.1111/1744-7917.12760

Lioy S, Carisio L, Manino A, Porporato M (2023) Climatic Niche Differentiation between the Invasive Hornet *Vespa velutina nigrithorax* and Two Native Hornets in Europe, *Vespa crabro* and *Vespa orientalis*. Diversity, 15, 495. https://doi.org/10.3390/d15040495

Lui Z, Chen S, Zhou Y, Xie C, Zhu B, Zhu H, Liu S, Wang W, Chen H, Ji Y (2015) Deciphering the venomic transcriptome of killer-wasp *Vespa velutina*. Scientific Reports 5: 9454 DOI: 10.1038/srep09454

Liu Y, Shu H, Long Y et al. (2022) Development and internal validation of aWasp Sting Severity Score to assess severity and indicate blood purification in persons with Asian wasp stings. Clin Kidney J 15: 320–327

Macià FX, Mechetti M, Corbella C, Grajera J & Vila R (2019) Exploitation of the invasive Asian hornet *Vespa velutina* by the European Honey Buzzard *Pernis apivorus*. Bird Study, DOI: https://doi.org/10.1080/00063657.2019.1660304

References

Makino S and Yamane, S (1980) Heat production by the foundress of *Vespa simillima*, with description of its embryo nest (Hymenoptera: Vespidae). Insecta Matsumurana, 19: 89-101

Makino S, Kawashima M, Kosaka H (2011) First record of occurrence of *Xenos moutoni* (Strepsiptera: Stylopidae), an important parasite of hornets (Hymenoptera: Vespidae: *Vespa*), in Korea. Journal of Asia-Pacific Entomology 14(1) 137-139. https://doi.org/10.1016/j.aspen.2010.09.001

Meiborg AB, Faber NR, Taylor BA, Harpur BA, Gorjanc G. (2023) The suppressive potential of a gene drive in populations of invasive social wasps is currently limited. Scientific Reports. Jan 30;13(1):1640.

Marris G, Brown M, Cuthbertson AG (2011) GB non-native organism risk assessment for *Vespa velutina nigrithorax*. http//www.nonnativespecies. org. (Accessed 12 Jan 2019)

Martin SJ (1990) Nest thermoregulation in *Vespa simillima, V. tropica* and *V. analis*. Ecological Entomology 15: 301-310

Martin SJ (1990b) Usurpation by conspecific queens of *Vespa simillima xanthoptera* Cameron (Hym., Vespidae). Entomologist's Monthly Magazine, vol. 126, 22-24.

Martin SJ (1992) Colony Failure in the Social Wasp *Vespa affinis* (Hymenoptera, Vespidae). Jpn. J. Ent., 60 (4):671-681.

Martin SJ (1995) Hornets of Malaysia. Malayan Nature Journal 49: 71-82

Martin SJ (2017) The Asian Hornet — threats, biology and expansion. IBRA and Northern Bee Books, Hebden Bridge, England

Matsuura M, Yamane S (1990) Biology of vespine wasps. Springer-Verlag, Berlin

Meurgey F, Perrocheau R (2015) Les Sarracénies piéges pour le Frelon à pattes jaunes. Insectes no. 177 (2) (in French)

Milanesio D, Saccani M, Maggiora R, Laurino D, Porporato M (2016) Design of an harmonic radar for the tracking of the Asian yellow-legged hornet. Ecology and Evolution 6 (7)

Mollet T, De la Torre C (2006) *Vespa velutina* — the Asian hornet. Bulletin Technique Apicole 33(4), 203-208. Translation in Bee Craft September 2007, 11-14

Monceau K, Bonnard O, Thiéry D (2012) Chasing the queens of the alien predator of honeybees: a water drop in the invasiveness ocean. Open J Ecol 2:183–191

Monceau K, Maher N, Bonnard O, Thiéry D (2013a) Predation pressure dynamics study of the recently introduced honeybee killer *Vespa velutina*: learning from the enemy. Apidologie 44: 209-221

Monceau K, Bonnard O, Thiéry D (2013b) Relationship between the age of *Vespa velutina* workers and their defensive behaviour established from colonies maintained in the laboratory. Insect Soc 60: 437–444

Monceau K, Arca M, Leprêtre L, Mougel F, Bonnard O, Silvain J-F, Maher N, Arnold G, Thiéry D (2013c) Native prey and invasive predator patterns of foraging activity: the case of the yellow-legged hornet predation at European honeybee hives. PLoS One 8 (6)

Monceau K, Bonnard O, Thiéry D. (2014a) *Vespa velutina*: a new invasive predator of honeybees in Europe. J Pest Sci 87:1–16

Monceau K, Bonnard O, Moreau J, Thiéry D (2014b) Spatial distribution of *Vespa velutina* individuals hunting at domestic honeybee hives: heterogeneity at a local scale. Insect Science 21, 765-774

Monceau K, Maher N, Bonnard O, Thiéry D (2015) Evaluation of competition between a native and an invasive hornet species: Do seasonal phenologies overlap? Bulletin of Entomological Research 105, 462-469

Monceau K, Thiéry D (2016) *Vespa velutina*: current situation and perspectives. Atti Accademia Nazionale Italiana di Entomologia Anno LXIV 137-142

Monceau K, Thiéry D (2017) *Vespa velutina* nest distribution at a local scale: an eight-year survey of the invasive honeybee predator. Insect Science. 24: 663-674

Monceau K, Tourat A, Arca M, Bonnard O, Arnold G, Thiéry D (2017) Daily and seasonal extranidal behaviour variations in the yellow-legged hornet, *Vespa velutina* Lepeletier (Hymenoptera: Vespidae). J Insect Behaviour 30 (2) 220-230

Nave A, Fernandes J, Ferreira MA, Rato F, Godinho J (2022) A Vespa-Asiática, *Vespa velutina*, espécie exótica e invasora em Portugal. Vida Rural, abril 2022. (in Portuguese).

Neov B, Georgieva A, Shumkova R, Radoslavov G, Hristov P (2019) Biotic and Abiotic Factors Associated with Colonies Mortalities of Managed Honey Bee (*Apis mellifera*). Diversity 11: 237; doi:10.3390/d11120237

Onofre N, Inês Portugal e Castro M, Nave A, Godinho J (2022) The natural predators of the Asian hornet (*Vespa velutina*) and the contribution of the European Bee-eater (*Merops apiaster*) to its control. Silva Lusitana, 30(1): 1–16, 2022. https://doi.org/10.1051/silu/20223001001 (In Portuguese)

O'Shea-Wheller TA, Curtis RJ, Kennedy PJ et al. (2023) Quantifying the impact of an invasive hornet on *Bombus terrestris* colonies. Communications Biology 6: 990

O'Shea-Wheller TA, Corbett A, Osborne JL, Recker M, Kennedy PJ (2024) VespAI: a deep learning-based system for the detection of invasive hornets. Commun Biol 7, 354 (2024). https://doi.org/10.1038/s42003-024-05979-z

Pawlyszyn B (1992) Nest relocation in the British Hornet *Vespa crabro gribodoi* Bequaert (Hym., Vespidae). Entomologist's Monthly Magazine 128: 203-205

Pazos T, Álvarez-Figueiró P, Cortés-Vázquez JA, Jácome MA, Servia MJ. (2022) Of fears and budgets: Strategies of control in *Vespa velutina* invasion and lessons for best management practices. Environmental Management. Oct;70(4):605-17.

Pérez-de-Heredia I, Darrouzet E, Goldarazena A, Romón P, Iturrondobeitia J-C (2017) Differentiating between gynes and workers in the invasive hornet *Vespa velutina* (Hymenoptera, Vespidae) in Europe. Journal of Hymenoptera 60: 119-133

Perrard A, Haxaire J, Rortais A, Villemant C. (2009) Observations on the colony activity of the Asian hornet *Vespa velutina* Lepeletier 1836 (Hymenoptera: Vespidae: Vespinae) in France. Ann. soc. entomol. Fr. (n.s.) 45 (1) : 119-127

Perry CJ, Søvik E, Myerscough MR, Barron AB (2015) Rapid behavioral maturation accelerates failure of stressed honey bee colonies. PNAS 112: 3427-3432

Pisa LW, Amaral-Rogers V, Belzunces LP, Bonmatin J-M, Goulson D, Kreutzweiser D, Krupke C, Liess M, McField M, Morrissey C, Noome DA, Settele J, Simon-Delso N, Stark J, van der Sluijs, van Dyck H, Wiemers M (2014) Effects of neonicotinoids and fipronil on non-target invertebrates. Environ Sci Pollut Res 22 (1): 68–102

Poidatz J (2017) PhD Thesis. De la biologie des reproducteurs au comportement d'approvisionnement du nid,vers des pistes de biocontrôle du frelon asiatique *Vespa velutina* en France. Ecologie, Environnement. Université de Bordeaux (in French)

Poidatz J, Bressac C, Bonnard O, Thiéry D (2017) Delayed sexual maturity in males of *Vespa velutina*. Insect Science. 10.1111/1744-7917.12452.

Poidatz J, Monceau K, Bonnard O, Thiéry D (2018a) Activity rhythm and action range of workers of the invasive hornet predator of honeybees *Vespa velutina*, measured by radio frequency identification tags. Ecology and Evolution 1-11

Poidatz J, Plantey RL, Thiéry D (2018b) Indigenous strains of *Beauveria* and *Metharizium* as potential biological control agents against the invasive hornet *Vespa velutina*. Journal of Invertebrate Pathology 153: 180-185

Poidtaz J, Plantey RJL, Thiéry D (2019) *Beauveria bassiana* strain naturally parasitizing the bee predator *Vespa velutina* in France. Entomologia Generalis March 2019. In press.

Poidatz J, Lihoreau M, Thiery D (2022) Patterns of food transfer in yellow-legged hornet nests revealed by heavy metal tracers. Entomologia Generalis 42(5)

Poidatz J, Chiron G, Kennedy P, Osbourne J, Requier F (2023) Density of predating Asian hornets at hives disturbs the 3D flight performance of honey bees and decreases predation success. Ecology and Evolution. 2023;13:e9902. https://doi.org/10.1002/ece3.9902

Pretre G, Asturias JA, Lizaso MT and Tabar AI (2022). Dipeptidyl peptidase IV of the *Vespa velutina nigrithorax* venom is recognized as a relevant allergen. Annals of Allergy, Asthma & Immunology, 129(1), pp.101-105.

Prezoto F, Nascimento FS, Barbosa BC, Somavilla A (Editors) (2021) Neotropical Social Wasps, Basic and Applied Aspects, Springer Nature, Switzerland, https://doi.org/10.1007/978-3-030-53510-0

Rebollo S, Díaz-Aranda LM, Martín-Ávila JA, Hernández-García M, López-Rodríguez M, Monteagudo N, Fernández-Pereira JM. (2023) Assessment of the consumption of the exotic Asian hornet *Vespa velutina* by the European Honey Buzzard *Pernis apivorus* in southwestern Europe. Bird Study. 2023 Jul 3;70(3):136-50.

Renoux J, Morin E, Dardenne B. Analyse comparee de l'attractivite et de la selectivite de trois dispositifs de piegeage de *Vespa velutina nigrithorax*. https://www.blog-veto-pharma.com/comparatif-efficacite-dispostifs-piegeage-frelon-asiatique/ (In French)

Requier F, Rome Q, Chiron G, Decante D, Marion S, Menard M, Muller F, Villemant C, Henry M (2018) Predation of the invasive Asian hornet affects foraging activity and survival probability of honey bees in Western Europe. Journal of Pest Science. 1612-4766

Requier F, Rome Q, Villemant C, Henry M. (2020) A biodiversity-friendly method to mitigate the invasive Asian hornet's impact on European honey bees. Journal of Pest Science. Jan;93(1):1-9.

Requier F, Fournier A, Pointeau S, Rome Q, Courchamp F. (2023) Economic costs of the invasive yellow-legged hornet on honey bees. Science of the Total Environment. Nov 10;898:165576.

Reynaud L, Guérin-Lassous I (2016) Design of a force-based controlled mobility on aerial vehicles for pest management. Ad Hoc Networks, Elsevier, 53, pp.41 - 52. <10.1016/j.adhoc.2016.09.005>. <hal-01427874>

Richter MR (2000) Social wasp (Hymenoptera: Vespidae) foraging behavior. Annu. Rev. Entomol 45: 121-150

Robinet C, Suppo C and Darrouzet E (2017) Rapid spread of the invasive yellow-legged hornet in France: the role of human-mediated dispersal and the effects of control measures. J Appl Ecol, 54: 205-215

Robinson W (2013) *Apis cerana* swarms abscond to battle and elude hornets (*Vespa* spp.) in northern Thailand. Journal of Apicultural Research 52(3): 160-172 DOI 10.3896/IBRA.1.52.3.08

Rodríguez-Flores MS, Seijo-Rodríguez A, Escuredo O, Seijo-Coello MD (2019) Spreading of *Vespa velutina* in northwestern Spain: influence of elevation and meteorological factors and effect of bait trapping on target and non-target living organisms. Journal of Pest Science. Mar 15;92:557-65.

Rodríguez-Flores MS, Falcão SI, Escuredo O, Seijo MC, Vilas-Boas (2021) Chemical profile from the head of *Vespa velutina* and *V. crabro.* Apidologie, 52, 548–560. https://doi.org/10.1007/s13592-021-00842-0

Rojas-Nossa SV, Novoa N, Serrano A, Calviño-Cancela M (2018) Performance of baited traps used as control tools for the invasive hornet *Vespa velutina* and their impact on non-target insects. Apidologie 49: 872.

Rojas-Nossa SV and Calviño-Cancela M (2020) The invasive hornet *Vespa velutina* affects pollination of a wild plant through changes in abundance and behaviour of floral visitors. Biol Invasions 22: 2609–2618 https://doi.org/10.1007/s10530-020-02275-

Rojas-Nossa SV, Dasilva-Martins D, Mato S, Bartolomé C, Maside X, Garrido J (2022a) Effectiveness of electric harps in reducing *Vespa velutina* predation pressure and consequences for honey bee colony development. Pest Manag Sci 78: 5142–5149

Rojas-Nossa SV, Álvarez P, Garrido J and Calviño-Cancela M (2022b) Method for Nest Detection of the Yellow-Legged Hornet in High Density Areas. Front. Insect Sci. 2:851010. doi: 10.3389/finsc.2022.851010

Rojas-Nossa SV, O'Shea-Wheller TA, Poidatz J, Mato S, Osborne J, Garrido J (2023) Predator and pollinator? An invasive hornet alters the pollination dynamics of a native plant. Basic and Applied Ecology 71: 119-128

Rome Q, Muller F, Olivier G, Villemant C (2009) Bilan 2008 De l'invasion de *Vespa velutina* Lepeletier en France (Hymenoptera: Vespidae). Bulletin de la Société entomologique de France, 114 (3): 297-302 (in French with English summary)

Rome Q, Muller F, Théry T, Andrivot J, Haubois S, Rosenstiehl E, Villemant C (2011) Impact sur l'entomofaune des pièges à bière ou à jus de cirier utilisés dans la lutte contre le frelon asiatique. In: Barbançon, J-M, L'Hostis, M (eds). Journée Scientifique Apicole JSA, Arles, 11 février 2011. ONIRIS-FNOSAD, Nantes pp. 18-20 (in French)

Rome Q, Sourdeau C, Muller F, Villemant C (2013) Le piégeage du frelon asiatique *Vespa velutina nigrithorax.* Intérêts et dangers. Conference paper. Journées Nationales GTV - Nantes 2013 (in French)

Rome Q, Muller FJ, Touret-Alby A, Darrouzet E, Perrard A, Villemant C (2015) Caste differentiation and seasonal changes in *Vespa velutina* (Hym.: Vespidae) colonies in its introduced range. Journal of Applied Entomology. 2015;139(10):771-82

Rome Q, Perrard A, Muller F, Fontaine C, Quilès A, Zuccon D & Villemant C (2021): Not just honeybees: predatory habits of *Vespa velutina* (Hymenoptera: Vespidae) in France, Annales de la Société entomologique de France (N.S.), DOI: 10.1080/00379271.2020.1867005

Ruiz-Cristi I, Berville L, Darrouzet E. (2020) Characterizing thermal tolerance in the invasive yellow-legged hornet (*Vespa velutina nigrithorax*): The first step toward a green control method. PloS one. Oct 6;15(10):e0239742.

Sánchez O, Arias A (2021) All That Glitters Is Not Gold: The Other Insects That Fall into the Asian Yellow-Legged Hornet *Vespa velutina* 'Specific' Traps. Biology, 10, 448. https://doi.org/10.3390/biology10050448

Sánchez O, Castro L, Fueyo Á, Borrell YJ, Arias A (2024). Early Alarm on the First Occurrence of the Southern Giant Hornet *Vespa soror* du Buysson, 1905 (Vespidae) in Europe. Ecology and Evolution. 2024 Nov;14(11):e70502.

Sánchez-Bayo F, Wyckhuys KAG (2019) Worldwide decline of the entomofauna: a review of its drivers. Biological Conservation 232: 8-27

Sauvard D, Imbault V, Darrouzet E (2018) Flight capacities of yellow-legged hornet (*Vespa velutina nigrithorax,* Hymenoptera: Vespidae) workers from an invasive population in Europe. PLoS ONE 13(6)

References

Shah F, Shah T (1991) *Vespa velutina*, a serious pest of honey bees in Kashmir. Bee World 72, 161–164

Spradbery JP (1973) Wasps: an account of the biology and natural history of social and solitary wasps. University of Washington Press, Seattle

Stainton K, McGreig S, Conyers C, Ponting S, Butler L, Brown P, Jones EP (2023) Molecular Identification of Asian hornet *Vespa velutina nigrithorax* Prey from Larval Gut Contents: A Promising Method to Study the Diet of an Invasive Pest. Animals 13, 511. https://doi.org/10.3390/ani13030511

Tan K, Hu Z, Chen W, Wang Z, Wang Y, et al. (2013) Fearful foragers: honey bees tune colony and individual foraging to multi-predator presence and food quality. PLoS ONE 8(9): e75841. doi:10.1371/journal.pone.0075841

Tan K, Radloff SE, Li JJ, Hepburn HR, Yang MX, Zhang LJ, Neumann P (2014) Bee-hawking by the wasp, *Vespa velutina*, on the honeybees *Apis cerana* and *A. mellifera*. Naturwissenschaften 94: 469–472

Tan K, Dong S, Li X, Liu X, Wang C, Li J, et al. (2016) Honey Bee Inhibitory Signaling Is Tuned to Threat Severity and Can Act as a Colony Alarm Signal. PLoS Biol 14(3): e1002423. doi:10.1371/ journal.pbio.1002423

Thiéry D, Bonnard, O, Maher N, Poidatz J, Monceau K (2014) Comportement de prédation du frelon asiatique à pattes jaunes (*Vespa velutina*) et protection des ruches par différentes stratégies de piégeage. Conférence Ravageurs et insectes émergents invasif, AFPP 2014, Montpellier France. (in French with English Abstract)

Thiéry D, Bonnard O, Riquier L, De Revel G, Monceau K (2018) An alarm pheromone in the venom gland of *Vespa velutina*: evidence revisited from the european invasive population. Entomologia Generalis. Nov 1;38(2).

Thiéry D, Doblas-Bajo M, Tourrain Z, Le Provost G, Núñez-Pérez E. (2023) Electrical traps, so called harps, efficient and selective against *Vespa velutina* workers predating on hives. Entomologia Generalis. Sep 1;43(5).

Thiery D, Monceau K (2024) Twenty years of attempting to control the *Vespa velutina* invasion: will we win the battle?. Entomologia Generalis. 2024;44(3):479-80

Tison L, Franc C, Burkart L, Jactel H, Monceau K, de Revel G, Thiéry D (2023) Pesticide contamination in an intensive insect predator of honey bees. Environment International, 176: 107975, ISSN 0160-4120, https://doi.org/10.1016/j.envint.2023.107975

Troadec N (2015) Finistère. Un homme décède après deux piqures de frelon asiatique. https://www.ouest-france.fr/bretagne/finistere-il-decede-apres-deux-piqures-de-frelon-asiatique-3614000. (Accessed 26 January 2019). (in French)

Turchi L, Derijard B (2018) Options for the biological and physical control of *Vespa velutina nigrithorax* (Hym.: Vespidae) in Europe: a review. Journal of Applied Entomology 142: 553-562

Ueno T (2015) Flower-visiting by the invasive hornet *Vespa velutina nigrithorax* (Hymenoptera: Vespidae). International Journal of Chemical, Environmental & Biological Sciences (IJCEBS) Volume 3, Issue 6

Van Itterbeeck, Joost & Feng, Ying & Zhao, Min & Wang, Cheng-Ye & Tan, Ken & Saga, Tatsuya & Nonaka, Kenichi & Jung, Chuleui. (2021). Rearing techniques for hornets with emphasis on *Vespa velutina* (Hymenoptera: Vespidae): A review. Journal of Asia-Pacific Entomology. 24. 10.1016/j.aspen.2021.03.009.

Verdasca MJ, Godhino R, Rocha RG, Portocarrero M, Carvalheiro LG, Rebelo R, Rebelo H (2021) A metabarcoding tool to detect predation of the honeybee *Apis mellifera* and other wild insects by the invasive *Vespa velutina*. Journal of Pest Science 95: 997-1007

Verdasca MJ, Carvalheiro L, Gutierrez JA, Granadeiro JP, Rome Q, Puechmaille SJ, Rebelo R, Rebelo H. (2022) Contrasting patterns from two invasion fronts suggest a niche shift of an invasive predator of native bees. PeerJ. May 10;10:e13269.

Victorian Government (2010) Invasive Plants and Animals Policy Framework, DPI Victoria, Melbourne. 29pp. https://agriculture.vic.gov.au/__data/assets/pdf_file/0009/582255/Invasive-Plants-and-Animals-Framework-Sep-22.pdf.

Villemant, C, Haxaire, J, Streito, J-C (2006) Premier bilan de l'invasion de *Vespa velutina* Lepeletier en France (Hymenoptera, Vespidae). Bull. Soc. Entomol. Fr. 111, 535–538 (in French, summary in English)

Villemant C, Barbet-Massin M, Perrard A, Muller F, Gargominy O, Jiguet F and Rome Q (2011a) Predicting the invasion risk by the alien beehawking yellow-legged hornet *Vespa velutina nigrithorax* across Europe and other continents with niche models. Biological conservation doi:10.1016/j.biocon.2011.04.009

Villemant C, Muller F, Haubois S, Perrard A, Darrouzet E, Rome Q (2011b) Bilan des travaux (MNHN et IRBI) sur l'invasion en France de *Vespa velutina*, le frelon asiatique prédateur d'abeilles. In: BarbanCon J-M, L'Hostis M (eds). Journée Scientifique Apicole JSA, Arles, 11 février 2011. ONIRIS-FNOSAD, Nantes, 3-12 (in French)

Villemant C, Zuccon D, Rome Q, Muller F, Poinar G, Justine J-L (2015) Can parasites halt the invader? Mermithid nematodes parasitizing the yellow-legged Asian hornet in France. PeerJ 3:e947; DOI 10.7717/peerj.947

Wang ZW, Chen G, Tan K (2014) Both olfactory and visual cues promote the hornet *Vespa velutina* to locate its honeybee prey *Apis cerana*. Insect. Soc. 2014, 61, 67–70.

Wen P, Cheng Y, Dong S, Wang Z, Tan K, Nieh J (2017) The sex pheromone of a globally invasive honey bee predator, the Asian eusocial hornet, *Vespa velutina*. Scientific Reports volume 7, Article number: 12956

Wycke M-A, Perrocheau R, Darrouzet E (2018) *Sarracenia* carnivorous plants cannot serve as efficient biological control of the invasive hornet *Vespa velutina nigrithorax* in Europe. Rethinking Ecology 3: 41-50

Yamaguchi Y, Ugajin A, Utagawa S, Nishimura M, et al. (2018) Double-edged heat: honeybee participation in a hot defensive bee ball reduces life expectancy with an increased likelihood of engaging in future defense. Behavioral Ecology and Sociobiology (2018) 72: 123. https://doi.org/10.1007/s00265-018-2545-z

Yamasaki K, Takahashi R, Harada R, Matsuo Y, Nakamura M, Takahashi J (2019) Reproductive interference by alien hornet *Vespa velutina* threatens the native populations of *Vespa simillima* in Japan. The Science of Nature 106: 15. https://doi.org/10.1007/s00114-019-1609-x

Yanez O, Zheng H-Q, Hu F-L, Neuman P, Dietemann V (2012) A scientific note on Israeli acute paralysis virus infection of Eastern honeybee *Apis cerana* and vespine predator *Vespa velutina*. Apidologie 43: 587- 589

Yang D, Zhao H, Shi J, Xu X et al. (2019) Discovery of Aphid Lethal Paralysis Virus in *Vespa velutina* and *Apis cerana* in China. Insects, 10: 157

Zhang L, Liu F, Wang X-L, Wang P-H, Ma S-L, Yang Y, Ye W-G, Diao Q-Y and Dai P-L (2022) Midgut Bacterial Communities of *Vespa velutina* Lepeletier (Hymenoptera: Vespidae). Front. Ecol. Evol. 10:934054.doi: 10.3389/fevo.2022.934054

Acknowledgements

My experience in Jersey was the impetus to writing this book, so special thanks to the welcoming Jersey Asian Hornet Group there, especially Alastair Christie (Asian Hornet Co-ordinator for Jersey), Bob Hogge, John de Carteret and Chris Isaacs. Check out the Jersey Asian Hornet Group facebook page for fascinating photos and videos. Thanks also to Francis Russell and Damian Harris, Asian Hornet Co-ordinators on Guernsey.

I went hornet hunting again in Jersey in 2023, with my hornet-hunting friends: Helen Tworkowski, Kate Crozier and Angus Deuchar who have all given me massive support and encouragement with the book, and lots of photos.

South West beekeepers Simon O'Sullivan, Judith Norman, Gerry Stuart, Sue Baxter, Colin Lodge and Lynne Ingram have also supported me enormously along the way.

I'd also like to thank the scientists who I've had discussions with, especially Dr Peter Kennedy from Exeter University, Dr Ana Diéguez-Antón from the University of Vigo, Dr Eric Darrouzet from the University of Tours, Dr Quentin Rome (MNHN) and Dr Denis Thiéry (INRA).

In looking at how other countries have been dealing with the yellow-legged hornet, I have had the pleasure of connecting with those who have been dealing with the yellow-legged hornet up close, and were able to hand on their direct experience, especially Dominique Soete (Belgium), Michelle Smeets and Sebastiaan Claus (Netherlands), Lukas Seehausen (Switzerland), and Kevin Baughen and Richard Noel who both keep bees in different parts of France.

Thanks to Jonathan How, layout mentor; and all those who have allowed me to use their photos (you are credited by your photos) — I *really* appreciate being able to add such amazing images. Any unattributed photos or illustrations are the author's. Thanks too to Dr Peter Kennedy, Dr Ana Diéguez-Antón, Dr Robin Wootton and Dr Clive Betts, all entomologists who have commented on various drafts. Hen Curtis gave me a great workspace in the depths of winter, which is very much appreciated.

Finally, I promised my partner Paul that I wouldn't have such a deadline this time around, but somehow the second edition seems to have taken longer than the first! Thanks for your patience.

Index

www.ingramcontent.com/pod-product-compliance
Lightning Source LLC
Chambersburg PA
CBHW051313020426
42333CB00028B/3325